D1014549

CLIMATE CRASH

CLIMATE CRASH

Abrupt Climate Change and What It Means for Our Future

John D. Cox

Joseph Henry Press
Washington, D.C.

Joseph Henry Press • 500 Fifth Street, NW • Washington, DC 20001

The Joseph Henry Press, an imprint of the National Academy Press, was created with the goal of making books on science, technology, and health more widely available to professionals and the public. Joseph Henry was one of the founders of the National Academy of Sciences and a leader in early American science.

Library of Congress Cataloging-in-Publication Data

Cox, John D.
Climate crash : abrupt climate change and what it means for our future / By John D. Cox.
p. cm.
Includes bibliographical references and index.
ISBN 0-309-09312-0 (cloth)
1. Paleoclimatology. 2. Climatic changes—Environmental aspects.
I. Title.
QC884.C69 2002
551.7′9—dc22

2005002387

Cover design by Michele de la Menardiere; photo © Paul Barton/ CORBIS.

Printed in the United States of America.

To
John
and
Margaret

CONTENTS

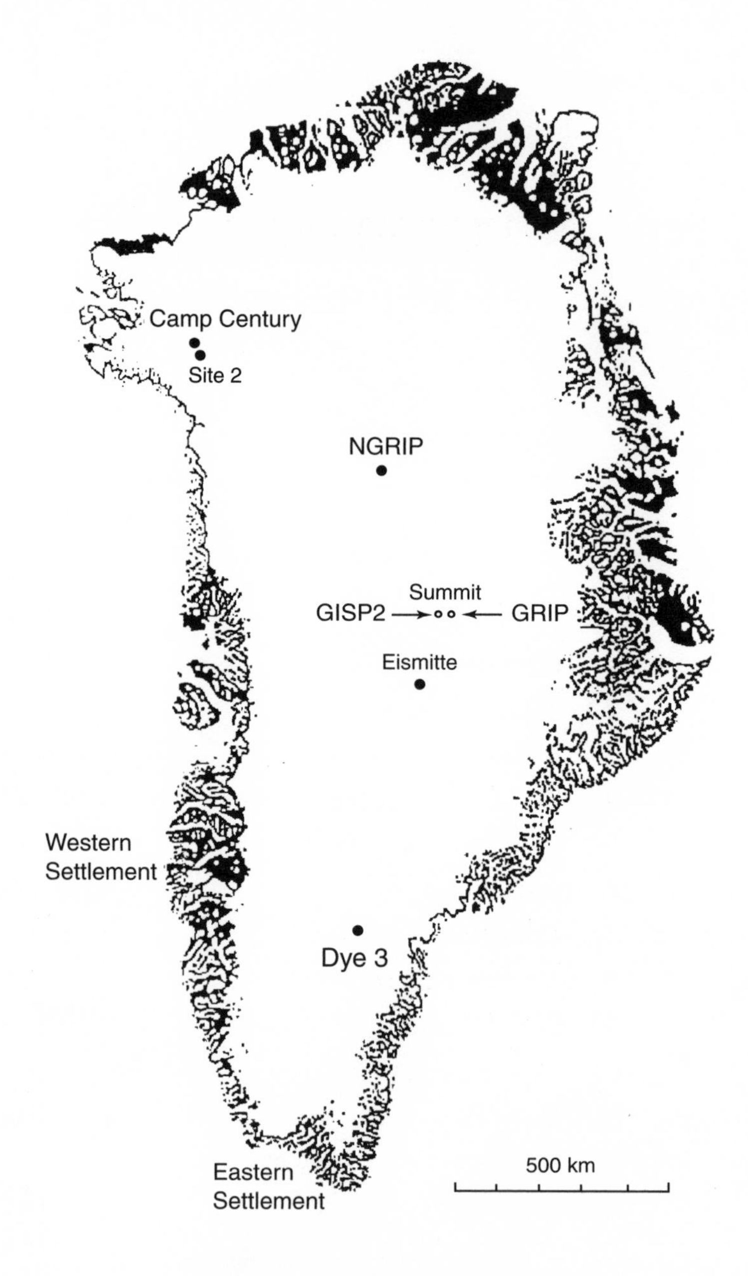

Greenland

Principal ice core drilling sites and Norse settlements

INTRODUCTION

These are exciting times in climate science. Discoveries that grew out of a line of research that began 50 years ago as a small geophysical field experiment are making their way onto the public stage. A big secret about how climate behaves was buried a mile deep in the polar ice on Greenland, and scientists went there and found it. What the earth was keeping from us was this: When change comes, it can be big and fast.

More than 110,000 years of snowfalls are piled up in a continuous layered sequence two miles deep atop Greenland. Scientists figured out how to core through this ice sheet as if it were an apple, and what they pulled up in the mid-1990s is the clearest record of climate that anyone has ever seen. Through some ingenious chemistry and other clever laboratory work, they are able to read this record like a book—to count individual years, to tell the temperature of the air when the snow fell, to estimate the amount of snow and determine the composition of the atmosphere at the time.

These cores tell a story about the last ice age that is nothing like the ponderous, slumbering epoch of cold stability that everyone has grown up hearing about. The whole record that we expected to be smooth as a knife blade is punctuated with enormous lurching changes between warm and cold, wet and dry. This is the signature of abrupt climate change—a result so surprising and contrary to some basic assumptions about climate that it is just now working its way through the earth sciences and beyond.

While this revolution in thinking is under way, climate and the science of its behavior are becoming topics of increasing public awareness. Signs of change are widespread, especially at high northern latitudes, where temperatures are rising faster than elsewhere on the planet and glaciers and sea ice are melting rapidly. Around the world, extreme weather events seem to be more common, and many people are on edge about what this portends.

Abrupt climate change is finding its way into the popular culture. Many people who may have been only vaguely aware that there were such beings as paleoclimatologists saw one portrayed as the hero of a blockbuster Hollywood motion picture in 2004. Who would have thought? Of course, 20th Century Fox won't win any awards for its rendering of the science of abrupt climate change in *The Day After Tomorrow*. Don't expect a single ice storm that fractures New York City, an outbreak of tornadoes that tear apart the skyscrapers of downtown Los Angeles, or a long-term ice age any time soon. Yet, though it doesn't happen in a day or two—that's the difference between weather and climate—it happens faster than you think. And one basic idea, unfortunately, Hollywood got just right: Climate change can be dangerous, even catastrophic.

While theater patrons were munching their popcorn, the Pentagon was mulling over a private think tank study it had commissioned that painted a particularly gloomy scenario of abrupt climate change as a threat to national security during the new century. "Military confrontation may be triggered by a desperate need for natural resources such as energy, food and water," its authors warned. This was not meant to be a prediction of the future, but rather a device to encourage strategists to begin thinking in different terms about climate.

But there is nothing imaginary about abrupt climate change or, for that matter, about this story of its discovery. It is not a hypothesis or a computer simulation. It is a solid theory supported by a careful reading of the remarkable direct evidence, the hard data that scientists pried from the earth itself—or, more exactly, the ice, sea, and land. In fact, other climate archives around the world not only confirm the Greenland record, they are yielding a very different picture than the previously supposed stability of the past 10,000 years, the period that saw the rise of human civilizations.

On top of this new view of a more changeable climate is the unnerving discovery that it is basically unpredictable. The climate is a chaotic system, like a stock market. All kinds of things are going on for different reasons, working on different timescales, and one day they line up together and the market comes tumbling down. If you educate yourself on the subject, and hedge your bets, you will probably be a more successful investor than if you don't—but you can never entirely eliminate the possibility that one day, to everyone's surprise, the market will crash. And even though we stand to learn a lot more about climate, this uncertainty may never go away.

Uncertainty poses a real problem for scientists trying to convey the idea that abrupt climate change might be a big risk for modern society. It poses an even bigger problem for policy makers trying to generate support for measures that might reduce those risks. When it comes to spending money or political capital, it is hard to attract a lot of interest in a patently unsure thing.

Of course, scientists familiar with the workings of the atmosphere have worried about the impact of industrial pollution for a long time. The brilliant meteorologist Carl G. Rossby, who ushered in the era of modern weather forecasting, warned in the 1950s about the potential of industrial gas pollution. "Nature can be vengeful," he said. Well, Rossby was taking a certain poetic license with that expression, suggesting that some immoral behavior was about to be punished; indeed the effects of abrupt climate change certainly may *feel* like vengefulness. For what it's worth, though, Rossby may have been right about the risks posed by changing the composition of the atmosphere. Climate scientists now realize that, just as a moving hand is more likely to throw a switch than a still one, *anything* that changes the system runs the risk of provoking abrupt climate change.

Climate science is young, and researchers speak of abrupt change with a real sense of discovery. Gerard Bond at Lamont-Doherty Earth Observatory, after analyzing seafloor cores that revealed abrupt change, saw the big changes coming in 1993. "It's like just before plate tectonics revolutionized geology," he told Richard A. Kerr of the journal *Science*. "Everything then was in a state of confusion. A few key pieces of evidence came to light, and then it looked simple. There could be a new theory coming out of this for how the earth's climate system operates."

This book does not present a grand new theory of how Earth's climate system operates, although it tells a remarkable story of researchers whose work has dramatically advanced their science toward that goal. Abrupt climate change is a brilliant new insight into the way the world works. In the end, there is still uncertainty. It is the nature of the climate system. Scientists don't have all of the answers, of course; science doesn't work that way. But they have more of them, and they are fascinating, and there is less confusion.

1

EISMITTE

Discovering the window to the past

Alfred Lohar Wegener, the German meteorologist who launched a revolution in the earth sciences, observed his fiftieth birthday in a small cave excavated from the accumulation of old snow high up on the inland ice of Greenland. He celebrated with an apple and some sweets. More important to him that day were the desperate circumstances facing the five men in the cave and the critical turning point of their scientific expedition. Outside the cave, a powerful, dangerously cold wind was blowing across the glacier. The long Arctic winter was closing in. The weather had been terrible for a month. Some 250 miles from their base camp, five men were crowded into a freezing vault provisioned for the winter with barely enough food for three. Nearly 10,000 feet up on the ice cap, the "Mid-Ice Camp"—*Eismitte,* in German—was without a radio transmitter, completely cut off from the outside world.

It was November 1, 1930, long before Wegener would be recognized as one of the great visionaries of science. A thoughtful and quietly attractive personality, Wegener inspired deep affection and respect from students and colleagues and most especially from the 20 countrymen who had followed him to Greenland that spring. At the same time, among many who knew him not at all, he was the object of a kind of loathing that only rarely is seen in science. In those days, more than a few geologists in Europe and the United States would have been perfectly happy to hear that this meddlesome fellow with his cockeyed theory about drifting continents was safely packed away in the frozen reaches of one of the most remote places on Earth.

Important as the occasion was to the history of geology, the events in Greenland are enriched all the more by what they would come to mean to the progress of climate science that just now is unfolding. Another important idea was about to be born. The sturdy and brave men of the Wegener Expedition, as it came to be called, were about to open the first window onto the finely detailed landscape of climates past that eventually would be illuminated by the Greenland ice. One of the men in the cave was about to dig down and look closely at the layers in the snow—to unlatch that window and peer out. Wegener would not live to witness this accomplishment, just as he would not live to see his profound insight into the terrestrial history of Earth so completely win over the science of geology 30 years later.

Since he first published his theory in Germany in 1912, Wegener's vision of a planet composed of continents drifting like icebergs on a sea of heavier crust had been nothing but trouble. His own father-in-law, the climatologist Wladimir Köppen, had urged him to give it up. The president of the American Philosophical Society had dismissed the idea as "utter, damned rot," and debate among geologists had been going on for a long time.

The continents seem to fit together like pieces of a jigsaw puzzle, Wegener argued. Similarities among fossilized plants and animals and geologic formations on land masses separated by oceans suggested a time when the lands of Earth were composed of a single large continent. The big question was how: What forces could cause such a supercontinent to fracture and its pieces to drift apart over the eons? The explanation of the mechanisms of plate tectonics and the turning over of the planet's viscous interior would revolutionize the earth sciences in the 1960s. However, neither Wegener nor anyone else had a satisfactory explanation at the time.

Established theory held fast. Men had spent years developing a theory of the molten history of Earth, proposing that the mountains, valleys, and oceans were formed during a long period of planetary cooling. In the process, Earth took on the look of a shriveling fruit, its skin cracking and folding as it shrank. To the leaders of geology, the configuration of the continents was immutable.

In the early days of the century, when the lines between the disciplines were more freshly drawn, many authorities took great personal offense that this outsider to their sciences would presume to contradict the received theories.

As a young man, Wegener had visualized the idea of continental drift as a singular truth and thereafter devoted much of his persuasive skills to marshaling the various sciences to the cause of establishing its substance. He challenged botanists, biologists, and geologists to abandon cherished beliefs and to think across the disciplinary lines they had worked hard to protect. This comprehensive approach eventually would be seen as vital to understanding the workings of Earth, especially its climate, but not in Wegener's day. The farther the young German ranged in pursuit of his theory, the more professional and personal antagonism he seemed to provoke.

It wasn't long before the established authorities closed ranks against Wegener's concept as if they were stamping out a plague. He never understood the depth of resistance to his thinking. Continental drift was received not merely as a mistaken idea but as an evil that jeopardized the credibility of geology as a science and the professional reputation of anyone who espoused it. Wegener was denied professorships at German universities, but eventually he found himself at the University of Graz in Austria. Ironically, though, in 1928, he was asked to lead a *German* expedition to Greenland.

The geophysicist Emil Wieckert at Göttingen University had developed the "seismic" method of measuring Earth's interior, setting off "artificial earthquakes" by means of explosions and measuring their waves. Wilhelm Meinardus, an influential Göttingen geographer and himself an Arctic specialist, proposed that Wegener lead an expedition that would field-test Wieckert's new method on the mysteries of the Greenland ice sheet.

At age 50, Wegener was highly regarded as a meteorologist. A popular lecturer, he was the author of an important and successful textbook, *The Thermodynamics of the Atmosphere*, which contained valuable insights into the formation of rain. Also, Alfred Wegener knew Greenland better than just about anybody in Europe. At the age of 26 he had served as meteorologist for a Danish expedition there in 1906-1908, becoming the first scientist to use kites and tethered balloons to study polar air circulation. During a second expedition in 1912, Wegener and a Danish companion became the first Europeans to spend the winter in the inland ice, and they trekked more than 600 miles across the great ice sheet.

During World War I and the following years of economic hardship, when scientific funding in Germany was scarce, Wegener could only dream of returning to Greenland. But in 1928 he persuaded the German sponsors to expand the expedition's scientific mission to include questions of meteorology, such as storm tracks and the climate of the ice cap, that had been raised during his earlier explorations of the ice sheet.

By November 1, 1930, as the five men in the ice cave sat down to what one described as a "war council," their ambitious scientific mission seemed very much in jeopardy. The mid-ice camp was the center of a three-station cross section of the ice cap that would make meteorological observations from coast to coast along the latitude of 71° North. This is where the critical questions about the depth of the inland ice, its climate, and its effects on regional weather would be investigated most effectively.

The transect of observing stations would test the theory put forward by William H. Hobbs, a prominent University of Michigan geologist, that the large ice sheets of Greenland and Antarctica create their own climates that are dominated by a "perpetual anticyclone" of high pressure and offshore winds. Among those most interested in the Hobbs theory and its promise of clear skies were executives and pilots of the fledgling commercial airline industry who were looking for the best routes across the North Atlantic. The meteorology for both regions turned out to be more complex than was thought at the time, although the subject would be a matter of debate for years as theorists tried to reconcile contrary field observations. In the case of Greenland, in fact, the frequent circumstances of low pressure, and of summer and winter snowstorms, would make the ice sheet a high-fidelity archive of climates past.

Anticipating the clear skies associated with high pressure, the Wegener Expedition was surprised by the frequency and the incredible severity of storms over the ice sheet. As one member of the expedition would report, "On the whole the weather in the central station was far worse than we had expected. Frequently we were under the influence of depressions with snowstorms and overcast skies."

So severe was the weather that the expedition was unable to transport adequate provisions and equipment 250 miles over the high glacier from the western station to make the mid-ice camp ready for winter habitation. Two motor-driven sleds powered by aircraft propellers had proven unreliable over the uneven surface of the ice. For the traditional dogsleds, the weather had been terribly hostile—storm after storm of soft, deep snow and day after day of powerful, bitterly cold wind. Already, by late October, temperatures had fallen as low as –65°F. Several supply missions had turned back, leaving the mid-ice camp without such basic items as a hut to live in and an adequate fuel supply.

Wegener had arrived just two days earlier on a last, desperate dogsled journey from base camp. For 40 days, he, fellow meteorologist Fritz Loewe, and the young Greenlander guide Rasmus Villumsen had slogged through terrible storms and cold. Most of their crew had turned back. What kept Wegener going was his fear for the lives of the two men he had stationed up on the inland ice and his concern that the expedition's scientific goals were about to be lost. Pushing on, they were forced to abandon the petroleum and scientific equipment intended for the camp. About all Wegener had accomplished bringing to the outpost was another mouth to feed: Loewe's feet were so seriously frostbitten he would be unable to make the return journey.

If the science mission were to be salvaged, the expedition leader knew he had to rely on the courage and fortitude of two men—Johannes Georgi, 41, a meteorologist from the Naval Observatory in Hamburg, who would make balloon-borne and other weather observations, and Ernst Sorge, 31, a secondary school geography teacher from Berlin, who would be responsible for measuring the depth and structure of the glacier.

At the war council, Wegener was happy and fresh, acting as if he had just come in from an invigorating walk. As Sorge would recall, "He was fired with enthusiasm and ready to tackle anything." If Sorge and Georgi were intent on abandoning the winter station, Wegener said he was prepared to stay there himself through the winter with Loewe. But the two men calculated that they had just enough food and fuel to keep three men alive for six months. In place of a simple hut, they had made a home in the ice. Sorge and Georgi assured their leader that they were comfortable in the ice cave and ready to carry on the scientific work. "Whatever may happen, the cause must not suffer," Wegener had written earlier to Georgi. "It is the sacred thing which binds us together. It must be held aloft under all circumstances, however great the sacrifices may be. That is, if you like to call it so, my expeditionary religion. It guarantees above all expeditions without regrets!" Wegener and the Greenlander set off early on the morning of November 1 to return to the base camp.

For the men who spent the winter of 1930-1931 atop the ice cap of Greenland, mere survival might have been achievement enough. Certainly it was for Fritz Loewe, who would endure incredible pain and suffering through the long winter. Yet Sorge and Georgi, fired with Wegener's expeditionary religion, did more. The Wegener Expedition would be famous among European weather and ice specialists of the age for the remarkable science it accomplished.

The highest tribute would come from another hero, the explorer Augustine Courtauld, who was a member of the British Arctic Air-Route Expedition, which, coincidentally, also lost its charismatic leader, Gino Watkins, that same year in Greenland. Courtauld had spent five months alone the same winter in a better-equipped ice cap camp 440 kilometers northeast of Eismitte. Snowdrifts had buried him alive, trapping Courtauld in his tent for days before Watkins arrived to dig him out. "Few people can realize what difficulties these men had to work under," Courtauld told a meeting of the Royal Geographical Society in London, on a December afternoon in 1932 when Sorge presented his findings. "They had neither house nor tent, very little fuel, and a bare sufficiency of food; yet they were not content with eking out their existence, but, as we have seen, they made every conceivable sort of observation under conditions which one would have thought would have entirely prohibited any sort of scientific work."

Sorge would write, "The snow-line was the line of demarcation between our labours. Everything above that was investigated by Georgi, everything below it belonged to my sphere." While Georgi braved the cold and wind to maintain a routine of meteorological observations, Sorge dug down into the ice, excavating a shaft like a hard-rock miner in a deep freeze. This first scientific effort to explore the interior of the Greenland ice was an especially dark, cold, lonely, and dangerous occupation in the winter of 1930. Sorge had to work alone. Georgi was busy with his own observations, and through the winter Loewe was too weak and wounded to get out of his sleeping bag. All Sorge had to work with was a saw, an ax, and a shovel. His only light came from a small kerosene lamp he made himself. He had no rope to haul the heavy chunks of ice out of the shaft or to fashion a ladder to drop down into it. Improvising, he dug a stair-stepped shaft that extended 35 feet below the ice cave, followed by a vertical shaft that dropped down another 15 feet, carrying out the big icy chunks by hand. One misstep on the dark stairway of ice could mean serious injury. At any time, he could be killed by a cave-in.

Work in the shaft occupied every afternoon. At an elevation of

9,850 feet, every exertion came at a high physical cost. On an average day, Sorge would haul out 300 pounds of ice before retiring to his sleeping bag, exhausted and cold. Using a ski pole and the brass leg of some instrument that had not arrived complete, he bore holes into the ice to monitor temperatures at various depths. Plugging the boreholes with sackcloth and old tin cans, and checking the thermometers every other afternoon, he was able to observe how deeply the warming effect of the previous summer had penetrated the ice. Diminishing with depth, of course, the warming advanced about a yard a month, he estimated, but no seasonal variations ever reached the deepest borehole 54 feet below the surface, where the temperature was always –19.57° F.

Working in the relative comfort of the cave's main room, Sorge would use the saw to carefully shape squared blocks of icy, granular old snow, or *firn*, and then measure their dimensions and weigh them. "In this way I was able to get a complete series of figures for the density of the firn at various depths," he wrote. "The density of course increased with depth." This pattern to the density intrigued him. Sorge was able to see layers in the icy snow, but they were not the kind of layers he was used to seeing in the glaciers of the Alps, where the warming sun of summer causes a dense formation of ice crust. On the ice cap of Greenland at 71° North, even summer temperatures were too cold to allow melting of the surface snow. In fact, Sorge told the geographers in London, "it is impossible to see the stratification caused by the change of summer and winter, as we can in the alpine glaciers." From 120 measurements taken during the winter, however, he was able to report that "the change of summer and winter layers was ascertained by changing densities. On the whole the density increases, but with small oscillations, so that each winter layer is a little denser than the adjoining summer layers. That is the reverse of the conditions in the Alps."

What most impressed fellow scientists at the time were Sorge's seismic measurements of the thickness of the inland ice. By setting off explosions at the surface, Sorge was able to register the waves reflecting off the bottom of the ice cap using his seismograph. By precisely timing the arrival of the waves, Sorge estimated that the ice sheet under the mid-ice camp was 6,500 feet thick, perhaps even as much as 8,775 feet. Could the Greenland ice sheet possibly be two miles deep? Presenting his results in London in 1932, Sorge heard the British geographer Frank Debenham call the thickness of the Greenland ice cap a "tremendous surprise" to glacier specialists. Debenham called it "the most momentous piece of work of a scientific kind connected with ice that has happened for a very long time. If, for instance, the height of the Ice Cap in the centre of Greenland is nearly 10,000 feet, while the height of the land underneath is only 1,000 feet above sea-level, it will at once lead to a large number of other questions with regard to the elasticity of the Earth's crust; as to what may have happened in the Pleistocene Ice Age, for instance, when we had presumably the same kind of ice-cap in Great Britain; in fact, there is no end to the echoes of these explosions of Dr. Sorge's."

Augustine Courtauld described the results as "entirely new to exploration" and found it "fantastic to hear that the depth of 9,000 feet of ice can be sounded by the echo from a trivial explosion of a few pounds of dynamite on the surface."

It would be some years before researchers would fully appreciate the value of Ernst Sorge's cold and lonely work in the winter of 1930 when he sawed and chipped and shoveled a shaft 54 feet down into the Greenland ice. Glaciologists at the time thought of the ice sheets almost exclusively as geological features whose movement across the landscape chronicled the slow waltz of a changing climate. Sorge's work led the way to the recognition of the ice itself as a remarkably faithful archive of atmospheric history. Among the first to recognize and describe the ice as a record of ancient atmospheres was the Swiss glaciologist Henri Bader, who saw polar ice as "a treasure trove" for scientists and in the 1950s first promoted the idea of drilling ice cores into polar glaciers. Under such ice sheets, he said, "every snowfall, including everything that fell with it, is, so to say, separately and safely filed for future reference." He described Sorge as "the first man literally to dig into the files."

At a symposium marking the beginning of the International Geophysical Year in 1957, Bader credited Sorge with what he called "the most fruitful law of polar glaciology." At any given location in the dry-snow region of a glacier, according to Sorge's law, the density of the snow in relation to its depth does not change with time unless the climate changes. Analyzing Sorge's data, Bader developed a mathematical expression that came to be known as "Sorge's Law of Densification." With this law as their guide, later investigators, such as the American Chester C. Langway, could determine the amount of annual accumulation even though they could not detect annual layers in their ice cores.

For all the science it accomplished, the 1930 Wegener Expedition would be most remembered for the singular fact that after Alfred Wegener departed Eismitte on the morning of November 1, 1930, to return to the base camp, he was never again seen alive. All through the long months of winter, nobody knew his fate. The scientists at Eismitte had no way of communicating to the base camp that Wegener had departed the mid-ice camp. Sorge, Georgi, and Loewe hoped and prayed that Wegener and Villumsen had successfully returned to the western coast. Under the circumstances, the scientists at base camp could only assume that their leader had decided to spend the winter at the mid-ice camp. It was not until May 12 that the truth was discovered. Wegener's body was found under several feet of snow 118 miles inland, about halfway between Eismitte and the base camp, the spot marked by two skis. His body had been carefully sewn up in two sleeping bag covers by his faithful companion, the Greenlander Rasmus Villumsen, who vanished on the inland ice. Sorge and others surmised that Wegener's heart had failed in the high-elevation exertion and extreme cold. His body was left in the Greenland ice cap.

A year later, in 1932, the widow Else Wegener wrote of her husband: "Now he lies in the land to whose exploration he devoted so many years of his life, a land to which he was ever attracted both by its scientific problems and the grandeur of the natural surroundings, in which he alone can live who is prepared to risk all else for the sake of self-preservation. Above his own safety he ranked that of his comrades, and when that was assured, his scientific work. He would not stay where he could have spent the winter in comparative security but where his scientific plans could not be carried out. He went out into the winter night and succumbed to its forces. But his death is consecrated by the nobility of his aims."

More than 30 years later—in the 1960s—scientists confirmed Alfred Wegener's theory of continental drift. It would be 30 years more—the 1990s—before the scientific contributions of Ernst Sorge were fully recognized with another revolutionary concept: abrupt climate change.

Sorge probably would not have been surprised by the pace of events. Reviewing his findings for the Geographical Society in 1932, he observed that, contrary to his own expectations, the expedition had not "mastered the main difficulties" in measuring and understanding the great ice cap. "It seems to me," he said, "that for a long time to come this will be the experience of all Greenland expeditions: results that have been attained will be the source of new problems."

2

YOUNGER DRYAS

The climate crash that changed the course of humanity

The ice was always melting back. Rivers were getting wider and flowing over new lands. Warm times were lasting longer. Forests were growing larger and richer, as were the big grasslands and their wild grains. Some 12,000 years ago, hunters were in the embrace of a stable, congenial climate stretching back farther than memory, perhaps farther than the stories told by the elders. For hundreds of years, the ice had been shrinking and a large swath of landscape had been bathing in steadily increasing warmth. In the fertile lands of West Asia, clans of an ancient people that would become known as the Natufians had long been accustomed to the good hunting of game and the easy gathering of fruits and wild grains. This was the way of the world for people 12,000 years ago. Nothing in the sky or the sea or on the land and nothing in memory warned them of what was to come.

The change in climate was a surprise of life-bending power, as if the whole world turned against the clans. Cold, dry winds began sweeping the countryside, withering every living thing in their paths. First the fruits were lost and then the grasslands, and eventually the forests were driven back. The rivers shrank and some became choked with advancing ice. Without warning or recognition, in just a few years, a colder, harder, shorter, and more difficult way of life set in. From beginning to end, as far as the Natufians were concerned, the time of wind and cold lasted forever. Like the old time of warmth and plenty, the hard new climate stretched out beyond memory, holding humanity in its grip for something like 470,000 days.

Some of the Natufian clans in western Asia responded to the change by migrating, as they always had done in the face of dwindling supplies of wild game and grains. Other clans adopted a new way of life. They settled in, cleared the land, and began to cultivate fields of einkorn, wheat, barley, and rye. The new way of life sustained the Natufians and spread among all the people occupying the fertile lands of the eastern Mediterranean.

Then, apparently just as suddenly, about 10,800 years ago, this cold, stingy climate finally gave way to a benevolent new pattern that enriched the countryside. In the small spare hamlets in the hills above the eastern Mediterranean, the sedentary clans would have noticed the increasing number of mild days and welcomed the more frequent rains. The woodlands thickened, the lakes and rivers enlarged, and soon the clans—farmers now—began sowing seeds of grain in the new wetlands. The land fattened, and before long the clans grew larger and so too did their villages. In the span of just a few years, as food became more plentiful, life among the clans became longer and more congenial, then soon more populous, and before long more elaborate.

What in the world happened? Why did the climate over much of Earth so suddenly reverse itself? What is it about the way of the world that makes such a thing possible? Why did this crippling cold, windy, dry regime hold sway over the sparsely populated lands of the Northern Hemisphere for 13 centuries?

Answers to these questions would be a long time coming. When the first evidence emerged from bogs in Scandinavia in the

1930s, no one had any idea what it meant, and in fact, researchers would puzzle over this abrupt climate change for the better part of the twentieth century. When earth scientists finally pried the answers from the ground, the students of human history, archeologists and anthropologists, began to recognize the change as a defining event in the civilizing of humanity—the sudden loss of wild food, provoking the transition from hunting and gathering to domestication and agriculture. By the end of the century, as the climate record came into sharper focus, a series of lower-magnitude climate shifts would emerge as transforming events in the rise and fall of civilizations.

Before scientists would be in a position to realize what had happened, however, they would have to invent new techniques for investigating the history and new ways of thinking about the character of climate. Fundamental mysteries of the comings and goings of ice ages were going to have to be solved, and the science itself would have to be reformed. Meanwhile, this catastrophic change of climate was given a perfectly obscure name and squirreled away in the literature of a science that was so new it didn't yet have a name.

The alpine plant *Dryas octopetala* thrives in conditions of climate and soil that most vegetation finds totally inhospitable. A delicate little white flower, *Dryas* is a mountain avens, of the family Rosaceae, a rose by another name. In the dry cold of Arctic and alpine mountainsides, it creeps along the rocks and clings to barren tundra clays that have been ground fine by glaciers. Treeless slopes that look over landscapes dominated by ice occasionally bloom with the white flowers of *Dryas octopetala.*

Early in the twentieth century, European botanists ingeniously fashioned their knowledge of plants—such as the cold tolerance of *Dryas octopetala*—into a powerful tool for the study of ancient climates. The work of pioneering botanists such as Knud Jessen at the University of Copenhagen was based on two critical findings: that a plant's individual pollen grains are shaped uniquely to its species, and that these pollen grains, although often appearing to be delicately structured, are virtually indestructible. Tiny pollen grains with their identifying features still intact could be found in soil deposits in which the leafy structure of their plants had long since decomposed. Jessen was among the first investigators to employ fossilized pollen grains as markers to trace the ancient advance and retreat of glaciers over northern Europe.

In the 1930s, when Jessen and others dug into the lakebeds and bogs of Scandinavia, an interesting pattern emerged. Soil differences were sharply defined, giving the sediments a layer-cake appearance. Dark, peaty bands were filled with the fossilized remnants of many plants that flourished in times of warm temperatures and abundant precipitation. Strips of light-colored, silty inorganic clays containing little more than the pollens of tundra-loving *Dryas octopetala* were signs of cold, dry times and the presence of glaciers. To distinguish between the bands of ice age clay, these pioneering researchers referred to the sequential appearances of the alpine flower. So the thickest, deepest, oldest band became the Oldest Dryas, and so on to the shallowest band of clay, laid down most recently, the climate shock that 12,000 years ago helped shape the course of human settlement and civilization—the Younger Dryas.

The explanations for this event and scores like it in recent geological history are remarkable, and to a surprising degree, they are key elements in modern descriptions of climate and its potential for change. At the turn of the twenty-first century, the effort to understand Earth's climate is one of the largest enterprises in science. It is the focus of hundreds of talented theorists, computer modelers, and dogged field investigators around the world. Great mysteries remain. They are different from those investigated by the Scandinavians as they dug into the bogs and lakebeds to sort out the pollens in the layered sediments, although in some respects the gaps in our knowledge are as wide now as they were then. As the eminent American climatologist Raymond T. Pierrehumbert recently observed, "Our understanding of climate is poor enough that if we did not have the paleoclimate data saying that the Younger Dryas did happen, we would never have anticipated anything like that."

Like research into many processes that left their evidence in the sediments of the earth, the study of ancient climates—paleoclimatology—grew up as a branch of the crusty science of geology. This practical circumstance critically shaped most thinking about climate throughout the twentieth century. Disparate processes such as the variations of climate, the evolution of species, the building of mountains, and the fluvial formation of valleys—all of these were understood to be different from one another, of course. But evidence for them came from the same material, was excavated by the same hands, and was interpreted at the time according to the same principles.

Central to the earth sciences is the concept of *geological time,* a scale that measures events in units of millions of years, and a certain assumption about the rate of geological change. Just about everyone who studies an earth science comes away with a lasting impression of this frame of reference. If Earth's 4.5 billion years is represented as a single day, humanity's time on the geological clock would be a fraction of the last minute. However valuable geological time is for studying Earth as a planet, the concept poses a particularly high barrier to the idea of abrupt climate change.

Long after they had come to terms with the mounting evidence that ice ages had come and gone several times in the past, researchers still resisted the idea that any of these changes had taken place rapidly. Even at their fastest rate, climate changes were calculated in thousands of years, units of time that measured the advance and retreat of glaciers, just about the last features of the climate system to respond to change. When the first evidence of the Younger Dryas emerged from the bogs of Scandinavia, mainstream geologists were thinking of glacier movements as local episodes of no particular global importance. They believed that the planet as a whole enjoyed a stable, "equable" climate.

The conventional view was expressed in 1906 by Scottish Professor John Walter Gregory as if it were the law of the land: "The first striking fact in the geological history of climate is that the present climate of the world has been maintained since the date of the earliest, unaltered, sedimentary deposits." Gregory's description of a "fairly constant" warm, moist atmosphere was entirely consistent with the prevailing geological thinking of the day—that Earth was a planet undergoing a very long, gradual process of cooling from a molten state. Geological processes such as mountain formation were the result of a slowly shriveling Earth, its crust wrinkling as it cooled, like the skin of a ripening fruit. The idea of ice ages waxing and waning, of ice sheets and glaciers moving back and forth over large areas of Earth, did not easily conform to such a notion.

Aside from volcanoes and earthquakes, themselves manifestations of timeless processes, nothing happens rapidly in that kind of geological world. And as long as the study of ancient climates was a child of geology, it was going to have to conform. For years to come, transitions between warmings and coolings were going to be accomplished at a respectable, uniformly gradual geological pace, over tens of thousands of years.

This question of the rate of change was no trivial matter to geologists. Nothing less than the integrity of their science was at stake. Then as now, life outside of the laboratory was full of people who were willing to believe a much wider range of wonderful stories about how the world works than mere science had to offer. Foremost among such believers were biblical fundamentalists with their supernatural catastrophes such as the purging flood wrought by an unhappy almighty God. Only after long rancorous debate had geologists in the nineteenth century overturned the notion of explaining what they saw in terms of the Great Flood and instead accepted an ice age past. There was no going back.

There were other stories, such as the fantastical pseudoscientific writings of Immanuel Velikovsky that were popular in the 1950s, competing with science. Velikovsky's cataclysmic vision of Earth's past shaped by colliding planets and asteroids is still kept alive by a cult of faithful followers.

Against this popular tide, geologists had staked a central principle: *The present is the key to the past. . . . No geological processes in the past happened any faster than any that happen at present.* This concept of uniformity meant that the natural physical laws are constant over time.

Outside of geology, one brave soul, a renowned weather scientist who was accustomed to studying events that transpired at a very different pace, was willing to speculate publicly that really rapid catastrophic climate change was not beyond the realm of possibility. In 1922, the British meteorologist C. E. P. Brooks suggested the past had seen "a series of startling changes of climate which almost merit the term 'Revolutions' of the old catastrophic geologists." How the very thought must have appalled geologists of the day! A few years later, Brooks went even further, suggesting that the atmosphere was subject to certain feedback processes that could provoke major change as rapidly as in "a single season."

Although these speculations by a weather scientist were not taken seriously by geologists of the day, the ground under the old line of thinking was continuing to shift as debate dragged on about ice ages, about their number and their geographical extent, and especially about the mechanisms that might cause them. Out of this dialogue emerged an idea that would prove to be critical to the discovery of abrupt change: The climate system has more than one mode of operation—a cold phase and a warm phase, at least—and somehow it switches between them.

Among the first to express this new vision of an unstable climate was the respected U.S. Weather Bureau physicist William J. Humphreys. In 1932, in an *Atlantic Monthly* article entitled "This Cold, Cold World," Humphreys suggested that "it is a scientific certainty that we are not wholly safe from such a world catastrophe" as another ice age. "Perhaps nothing of this kind will happen for a long, long while, but sooner or later it surely will if the future can be inferred from the past," he wrote. "When this will be we have no sure means of knowing, but we do know that the climatic gait of this our world is insecure and unsteady, teetering, indeed, on an ice age, however near or distant the inevitable fall."

At midcentury, speculations such as these were so far outside of conventional thinking on these subjects that they did not even arouse debate. During this era, the science of climatology grew up as an applied discipline that itself was heavily invested in the idea that climate was essentially unchanging. Climatologists were in the business of compiling the statistics of weather—tabulating the data recorded by instruments that had been in place less than a century—and calculating its dominant patterns according to region and season. This information, this picture of "normal," was especially important to engineers who were designing dams, hydroelectric power plants, and other large public works projects. However unsteady the climatic gait in the eyes of weather scientists, however inevitable the fall to a new ice age, climatologists and geologists continued to hold the climate system of the past to the slow, gradual pace of change they saw in the climate system of the present.

The Younger Dryas and the reality of abrupt climate change did not emerge easily from the early data. Wrinkles and spikes in the record were open to a wide range of interpretations—seldom were they taken to be meaningful climate transitions. Dramatic changes in sediment layers under lakes, for example, most often were ascribed to local events such as floods or fires or landslides. The science was young, and the data were full of unexplained upendings. In their searches for the big swaying rhythms of ice ages, geologists saw evidence of dramatic rapid change as "local noise" that had to be filtered out of their reports.

New ways of investigating the past were going to be invented, and new technologies—inspired by the exploration for oil—were going to be brought to the problem. The early fossil pollen research was founded on a train of thought that would spawn a whole range of illuminating studies: One piece of evidence can "stand in," or serve as a *proxy,* for another. That is, although researchers can't directly measure the changes in temperature and other features of past climates, they can detect them indirectly in other biological and geological evidence.

Proxy evidence is a primary method of most modern investigations of ancient climates, ingeniously employing all kinds of data found on land and ice and in the sediments under the lakes and oceans of the world. In addition to fossil pollens, researchers would discover reliable climate proxy information in the chemistry or unique structure of the tiny shells of marine plankton, the fossilized dung heaps of packrats, the carcasses of beetles, the annual bands of growth in ocean corals, the growth rings of trees, the chemical makeup of the green eggshells of emu, and most important to the discovery of abrupt climate change, of course, the composition of one of the purest substances on the planet—polar ice.

Still, in the first half of the twentieth century, the pollen records that seemed so obvious in Scandinavia were not as obvious in North America, and at the time no one had any way to reliably correlate the record of the past found in one region with what looked like climate changes in another. Among the first to make the attempt was the Swedish geologist Gerhard De Geer. At the turn of the century, De Geer began studying the coupled layers—or *varves*—of clay exposed by the excavations for building foundations during the expansion of Stockholm. He recognized a recurring pattern of differing grain coarseness in the varves and correctly surmised that they had been caused by seasonal changes, the coarser grains being deposited during annual spring surges of melt water from nearby glaciers. Using methods of *stratigraphic analysis* that would become common to later climate investigations, De Geer had discovered a calendar. These regular seasonal laminations allowed him to count the years back in time. He developed meticulous chronologies of the past 12,000 years in Sweden. In 1920, De Geer traveled to North America, where he studied sediment laminations in New England and thought he recognized features from his Swedish varves. He failed to persuade his colleagues, however, and his pioneering methods of analysis fell out of favor with a whole generation of geologists.

At the same time, in the American Southwest, examining layers of another kind, another innovative researcher was inventing the science of *dendrochronology* (from the Greek words *dendron* for tree and *chronologia*, log of time), the dating and study of annual growth rings of trees. It was Arizona astronomer Andrew Ellicott Douglass who first recognized that the differences in the widths of the growth rings of trees were related to food supply, to the availability of water, and to temperature and so were a record of climate change. Tree-ring researchers typically extract a core from a tree trunk, count and measure its annual growth rings, and determine the age of the wood by carbon-14 or other dating methods. A wide ring indicates plentiful water and good growing conditions and a narrow ring often indicates drought. Douglass realized that the best climate records came from trees growing in "water-stressed" areas that relied exclusively on rainfall or fog. Using ponderosa pine and then giant sequoia in northern California, he was able to develop overlapping sequences going back more than 3,000 years.

Applying his new science in the Four Corners region of the southwestern United States, Douglass matched the growth patterns in local trees to those found in the timbers used as beams of the elegant stone and adobe dwellings of the ancient Anasazi Indians, thereby determining the date of those ruins. After 500 years, the Anasazi Pueblo civilization suddenly collapsed in the thirteenth century during 26 years of drought.

In the 1930s, while the Scandinavians were doing their early pollen investigations, the German scientist Wolfgang Schott was launching a line of research using another instance of unique miniature architecture as a proxy of ancient temperatures. Examining meter-long ocean sediment cores taken by a German expedition in the Atlantic Ocean, Schott found that different layers were composed of the calcium carbonate skeletons of different populations of tiny drifting aquatic plants or animals, or *plankton*—some that thrive in warm waters, others that are more tolerant to cold.

The study of seafloor sediments would prove to be one of the richest lines of climate research, revealing the Younger Dryas as an abrupt change of widespread consequence as well as other rapid climate events. Later scientists would look beyond the different population assemblages of marine organizations on the seafloor and measure oxygen isotope ratios and other chemical and physical properties in the sediments to tease out more details of ancient climate history.

First, however, certain technical problems had to be overcome. The first mechanical process of extracting a core from the seafloor consisted of dropping through the depths a heavily weighted pipe that pierced the soft sediment like a cookie cutter. This method deformed the material, as friction between the corer and the ooze obscured the layers of the upper part of the core and limited the sample length to just a few meters. The problem was solved by a series of improvements, beginning in the late 1940s with the development of a coring device equipped with an internal piston that was raised as the corer was driven into the sediment, its suction effect counterbalancing the core-wall friction and allowing longer cores to be obtained. More fundamental problems were not so congenial to technological solutions, however. Over much of the ocean, the seafloor is inhabited by worms and other creatures that burrow through the sedimentary slime, blurring the lines between the layers that signify climate shifts. And in many areas of the open ocean, the rate of buildup of sedimentary deposits was simply too slow or uneven to leave a clear record of changes in ocean conditions. As these sediment investigations expanded, however, researchers found key places in the oceans where up to 200 meters of continuous, undisrupted, annually layered cores reveal thousands of years of climate change.

After World War II, the dawning Atomic Age began to shed new light on the Younger Dryas, on ice ages, and on the question of the pace of climate change. The ability to see more deeply into the structure of matter would have applications far beyond the military and the new weapons of mass destruction it made possible. Nuclear chemistry gave earth scientists two powerful new instruments: a clock and a thermometer.

In 1947, at the University of Chicago, chemist Willard F. Libby discovered a powerful new technology known as radiocarbon dating. Libby would win the Nobel Prize in Chemistry in 1960 for developing this geological clock. The technique relies on the ability of nuclear technology to measure the occurrence of different *isotopes* or slightly different atoms of an element. All carbon atoms contain six protons; most also contain six neutrons, giving the element an "atomic weight" of 12, which chemists and earth scientists speak of as "carbon-12" and write as ^{12}C. Some carbon atoms, however, contain seven neutrons (^{13}C) and others contain eight neutrons (^{14}C). Carbon isotopes with six or seven neutrons are stable, but ^{14}C, with eight, is unstable, or radioactive. This so-called radiocarbon is constantly emitting radiation as it transforms itself to a stable element, nitrogen-14. For any given quantity of atoms, this transformation is half complete every 5,730 years, a period referred to as the "half-life" of ^{14}C. Radiocarbon is constantly formed by cosmic rays bombarding the upper atmosphere, so all living organisms are always taking in a fresh supply of ^{14}C, along with the more abundant ^{12}C isotope, with the air they breath and the food they eat. The relative proportion of ^{14}C to ^{12}C remains roughly constant over time. When an organism dies, however, radiocarbon is no longer replenished but continues to decay. By measuring the ratio of ^{14}C to ^{12}C isotopes in a sample of dead organic material, therefore, scientists are able to determine how long ago it died.

Libby tested his new method on samples of wood from trees in the ancient Two Creeks Forest Bed of Wisconsin and found, unexpectedly, that the last big ice age surge in North America appeared to coincide with the Younger Dryas in Europe. Yale University geologist Richard Foster Flint used the new radiocarbon dating methods in 1955 to calculate that ice age glaciers had retreated from the Great Lakes region at the same rate as glaciers move in modern times. His conclusion reassured his colleagues: Past as present, the pace of events "rests on a sound uniformitarian basis." But uniformity, and the gradual pace of change that it implied, was beginning to lose its grip on climate science, and the new radiocarbon clock would hasten the day.

At the same time, a new level of detail of climate history would be revealed by the work of the geochemist Harold C. Urey, another Nobel laureate at the University of Chicago and another veteran of atomic bomb development. In 1947, in a famous paper in the *Journal of the Chemical Society* in London, "The Thermodynamic Properties of Isotopic Substances," Urey proposed that the ratio of stable isotopes of oxygen—the common ^{16}O and the rare

^{18}O—could be used as a "paleothermometer." Many scientists date the beginning of paleoclimatology to this paper. In particular, Urey told a meeting of the American Association for the Advancement of Science in 1948 that the ratio of the two oxygen isotopes in a common organic chemical was critically sensitive to temperature. "The temperature coefficient for the abundance of the oxygen isotope in calcium carbonate makes possible a new thermometer of great durability," he said.

At Urey's urging, the Italian geologist Cesare Emiliani measured the ratio of the two stable isotopes of oxygen—^{18}O and ^{16}O—in the skeletal calcium carbonate shells of foraminifera, tiny marine animals found in the sediment cores pulled up from the ocean depths. According to Urey's theory, the shells of the planktonic creatures took up more or less of the rare ^{18}O isotope according to the temperature of the seawater during their lives. Testing cores everywhere he found them, Emiliani produced a temperature record going back nearly 300,000 years. Researchers argued at length over the meaning of Emiliani's data. Urey was correct in theory, but the isotope differences caused by ocean temperatures were confounded by the physics of another process. Evaporation of seawater preferentially takes up ^{16}O, leaving behind an ocean enriched with ^{18}O during periods when ice sheets are building up over land. Eventually, researchers would conclude that Emiliani had captured the history not only of ocean temperatures but of the volume of ice on land as well—the waxing and waning of ice sheets.

A revolution was under way. Researchers would refine the methods of nuclear chemistry that tease time and temperature from ancient relics and extend their application. In particular, Danish geophysicist Willi Dansgaard would devise a method of using ratios of the oxygen isotopes in polar ice to calculate the temperature of the air when the ice fell as snow.

At the same time, a new generation of earth scientists seemed to be more open to new ideas about the pace of geological events. Out of the haze of prewar conjecture emerged a picture of increasingly fine detail, full of surprises: more and longer-lasting advances of ice sheets than anyone thought; sharper changes in temperature and precipitation than anyone supposed. The rate of change would be subject to continuing debate, but as researchers periodically reviewed the accumulating evidence, on each occasion the time horizon for climate transitions seemed to be shorter. Along the way came an increasingly clear vision of the Younger Dryas.

A young geochemist, Wallace S. Broecker, and other researchers at Lamont Geological Observatory in New York set to work applying the radiocarbon clock to a variety of ocean and lake sediments and other climate evidence, and their 1960 report introduced a new sense of the catastrophic potential of climate. "Evidence from a number of geographically isolated systems suggests that the warming of world-wide climate which occurred at the close of the Wisconsin glacial times was extremely abrupt," Broecker wrote. The Lamont team examined new evidence from sediments in the Atlantic and the Gulf of Mexico, from rain-fed lakes in the American West, from the Great Lakes Basin, and from pollen assemblages in Europe, and in each case "the radiocarbon age determinations suggest that the changes occurred in less than 1,000 years close to 11,000 years ago."

Were the data pointing to the end of the Younger Dryas? Using 1950s methods, Broecker couldn't be sure. The margins of error in radiocarbon dating still were being measured in centuries. Observing the pollen data from Europe, Broecker reported that he had found evidence of the Younger Dryas in the western lakebeds and in the Great Lakes—but not in the deep-sea cores or the Gulf of Mexico. "Certainly if the abruptness of such a major change in world climate can be firmly established it will be of the utmost importance that acceptable theories are able to account for it," he wrote. The comment was prescient. Broecker would devote much of his long, illustrious career to just such an undertaking, becoming the leading theorist on the subject.

As it happened, in the 1950s, a little-known, highly experimental project under way in Greenland would lead to the dawn of a whole new day in the study of ancient climates. Did succeeding years of polar weather leave a reliable record of climate in polar ice? Is it possible to drill into it and pull up a core and to read it as one can read the layers of ooze on the floors of oceans and lakes? These were the questions that a small group of researchers confronted in Greenland. The answers would lead to an astonishing new vision. A great corrective lens would be applied to the study of ancient climates, and it would be made of polar ice.

3

BEDROCK

Penetrating polar ice

The prospect of trekking up to Greenland, drilling into its ice sheet, and extracting a core to bring back and analyze in a freezing laboratory was an idea that, in the pre-World War II United States, attracted no one. Midwesterners know winter cold about as well as anybody, of course, but the country as a whole has no particular cultural affinity for the world of glaciers. European boys like Alfred Wegner may have grown up dreaming of heroic exploits on the polar ice, but the dreams of American youth were more likely to be set in warmer climes. Whatever territorial claims to the far north of Greenland that the polar expeditions of the American Robert Peary had established at the turn of the century were bartered away at the first opportunity in a telling transaction with Denmark in 1917. The United States ceded to the Danes all interests in northern Greenland and paid an additional $25 million in gold in exchange for the Virgin Islands. Norwegian hunters and fishermen kicked up a fuss when Denmark subsequently claimed Greenland as a colony, but there were not many complaints in the United States about the deal. Even as the importance of Greenland's weather to the growing trans-Atlantic aviation industry was becoming increasingly evident, among researchers in the United States the study of ice caps and glaciers was somebody else's science.

In 1940, German troops invaded Denmark and soon established a valuable radio aviation weather station on the east coast of Greenland. In 1941, the United States agreed to protect Denmark's sovereignty while using Greenland militarily to help secure the status quo in the Northern Hemisphere. The U.S. Coast Guard destroyed the German weather station in 1944, and the United States fought off several attempts by the Germans to regain a weather observation foothold in Greenland. Weather forecasts from Greenland were critical to the operations of Allied forces in the North Atlantic.

In 1947, the war over, Denmark asked that the 1941 Greenland agreement be rescinded. By then, however, the United States had laid claim to a new territorial interest in the region. Terms of an agreement that was finally reached in 1951 gave the United States the right to build an air force base at Thule and to member states of the North Atlantic Treaty Organization the right to use all military facilities on the island. In the 1950s, as the chill of the Cold War descended, the United States Army had a new adversary in mind and new national imperatives. It was intent on maintaining a strategic presence in Greenland to keep a close polar eye on the Soviet Union. Before long, the Distant Early Warning system formed a necklace of 63 radar stations along the 69th parallel from northwestern Alaska through central Greenland.

Pursuing a new interest in the geophysics of cold regions, the Army created a new unit within its Corps of Engineers. The Snow, Ice and Permafrost Research Establishment first set up shop at the University of Minnesota at St. Paul and then moved into its own special laboratory in Wilmette, Illinois, north of Chicago. The Pentagon's choice to develop the new science program was a young Swiss geologist, then teaching at Rutgers University in New Jersey, who had called the military's attention to the lack of cold regions research in the United States.

Henri Bader had spent the war years in various mining enterprises in South America and the West Indies, although he was best known in the United States as an authority on snow. A student of the eminent Swiss geologist Paul Niggli, Bader in 1938 had written the first chapter of *Snow and Its Metamorphism*, a famous work on the subject that the U.S. Army translated from the German and published in 1954. The War Department General Staff had sent Bader to Europe to survey the state of the art of cold regions research in England, France, Switzerland, and Germany, where ice and snow and alpine glaciers had been studied for years, and it was his report on the subject in 1947 that spearheaded the creation of the Army's new lab.

What most interested European researchers about alpine glaciers, aside from their inclination to collapse into dangerous avalanches, were the patterns they left on the landscape during their long-term advances and retreats in response to climate fluctuations. These telltale patterns and deposits, the moraines of rubble and the boulders left in odd places, were central to Louis Agassiz's successful argument for the existence of ice ages in Earth's history, one of the great achievements of nineteenth-century geology and a subject that continues to enthrall scientists. It was this line of geological thought—the grand movement of glaciers in response to long-term climate change—that became the frame of reference and the standard for the timescale of change for generations of earth scientists.

What most interested Henri Bader, however, was a very different line of thought that was first taken up by Ernst Sorge at Eismitte. Looking at glaciers through the eyes of a mineralogist, Bader wanted to learn everything there was to know about their actual content. His years of mining in Argentina and Colombia may have inspired this thinking of snow and ice as a special kind of ore that was worthy of very close examination. Bader traveled to Alaska as a member of a pioneering expedition that drilled 100 meters into the Taku Glacier and extracted a core. This early exploratory work in ice from a temperate region, the Juneau Icefield, wasn't exactly what he had in mind, but it seemed to prove the concept.

It was Henri Bader's audacious ambition, in the mid-1950s, to find a way to penetrate the ice caps of Greenland and Antarctica and investigate the composition of polar ice in ways that would reveal its finest details. This brilliant idea was inspired by a singular insight: that locked in the polar ice is a unique record of atmospheric conditions going far back in time.

Bader believed that polar ice contains information that is obscured in the ice of glaciers in warmer climates. Rain and heavy summer melting cause water to percolate deeply through the critical high-elevation "accumulation area" of glaciers in temperate climates, Bader observed, and so "washes out much of the detail of the record of past precipitation." In contrast, in the perpetual cold of polar climates, summer temperatures remain below freezing, so all precipitation falls as snow, depositing layers that remain frozen and "dry" and relatively unchanged. To Bader this was a vital difference. It meant that "the detail" was intact; the sequence of weather and climate events at high latitudes was preserved, each year layered upon the other, like the growth rings of a tree, a faithful record of climate history.

That was his theory, at least. But exactly what were these mysteries in the polar ice? Were they really worth solving? Was it even physically possible to drill into a polar ice cap and extract a core? And exactly how would one go about deciphering the information extracted? It was totally exploratory, and it looked very expensive and troublesome. No one had succeeded in boring into the polar ice at any great depth and, in fact, no one could say with any certainty in the mid-1950s that, mechanically speaking, it could be done. A critical and unlikely moment in the history of climate science was at hand.

Here was a European, an irascible, urbane alpine scientist, espousing a very European idea about the scientific value of studying glaciers—not their advances and retreats over geological time, like most glaciologists, but their actual *content.* Here he was, chief scientist of a research and engineering laboratory with the wherewithal of the American military and a national security mission to find out everything there was to know about cold regions.

To the Army brass in the 1950s, polar ice was most interesting as a kind of amorphous rock, a potential building material, as surfacing for landing strips and possibly bomb shelters. However, what interested the Army most about the ice cap was not what it might reveal but what it might hide. Could it hide intercontinental ballistic missiles? Could it hide tanks and even aircraft? Even with the demands of the Korean War being felt, the military's Cold Warriors were going to follow this line of thought as far as it would take them. The Army and its engineers wanted to know about the hardness of the ice, its tensile strength, and its deformation characteristics. Deep-time, deep-core ice research into ancient climates was not on their drawing boards.

So what in the world could bring together Bader's bold yet almost dreamy expectations that something scientifically valuable would come of deep-core drilling and the U.S. military's totally pragmatic determination to conquer the ice? As it happened, Bader had some remarkably good timing on his side.

The summer of 1957 would mark the beginning of a unique global scientific undertaking known as the International Geophysical Year, or IGY. Nations on both sides of the Iron Curtain somehow were managing to plan and execute an ambitious, coordinated research agenda that would enrich the physical sciences of Earth for years to come. Thousands of geologists, oceanographers, meteorologists, and other earth scientists were going to be dispatched around the world on a wide variety of field investigations. Among the major subjects of international research would be the mysterious continent of Antarctica.

As an influential member of the Committee on Polar Research of the National Academy of Sciences, Bader proposed that the United States sponsor a project to drill as deeply as physically possible into the Antarctic ice cap and extract an ice core of the interior of the polar glacier. To the U.S. committee appointed by the National Academy of Sciences to determine the American contributions to the IGY, Bader offered a tantalizing description of its possibilities.

"Two thirds of the area of the Greenland ice sheet and practically all of the Antarctic ice sheet are permanently dry," he noted, and temperatures are almost always subfreezing. All precipitation falls as snow. So every snowfall, including everything that fell with it—volcanic ash, meteorites, spores, and bacteria—is "separately and safely filed for future reference" by being buried under later snowfalls.

"The Greenland and Antarctic snow layers are a treasure trove for the scientist." Bader noted, and added that the treasure was not without military importance. In the ice was a record of the global atmosphere's response to the nuclear age since the first atomic detonations in 1945. Scientists monitoring radioactive fallout could go back several years into the "files" of polar ice in Greenland and Antarctica and measure some things they may have missed.

By analyzing snows back to preindustrial times, Bader also thought it would be possible to track atmospheric contamination by industrial activity.

Bader succeeded in persuading the U.S. National Committee to sponsor a project to extract two deep cores from the Antarctic ice sheet, at Marie Byrd Station in the interior and on the Ross Ice Shelf. Such an ambitious undertaking would require substantial preparation. A scientific and engineering crew would have to be sustained in a very remote and extreme environment for months at a time. Bader proposed a field test project in Greenland, where the U.S. military presence was already established. Roughnecks could experiment with drilling methods and equipment, and scientists could explore different techniques of ice core analysis. From a scientific point of view, the research program outlined by Henri Bader most logically would have fit within the purview of the U.S. Geological Survey, a venerable agency with a long history of earth science research. In the 1950s, however, some other secret logistical and strategic Cold War imperatives were at play. The U.S. committee accepted Bader's proposal that his own Army lab, the Snow, Ice and Permafrost Research Establishment (or SIPRE), would do the job.

In May 1956, 26-year-old Chester C. Langway Jr. of Worcester, Massachusetts, was hired as a civilian researcher at SIPRE's new Wilmette, Illinois, laboratory. Langway, who had served 15 months with the Army in Korea and 38 months with the Air Force in Germany, had a freshly minted master's degree in geology from Boston University. Like most of his geology classmates, Langway had thought he was on his way into the business of exploring for oil and gas when he heard about the job opening at the Army lab, but the geology of ice sounded more interesting. By June he was in Greenland.

Drilling already was under way at Site 2, a location in northwestern Greenland that was chosen for its logistical convenience 220 miles east of Thule Air Base. A nearby Army research facility and radar station added to the convenience of Site 2. In the field of deep-core polar ice sheet drilling, the summer of 1956 was the beginning, and Site 2 was ground zero. Eventually Langway would write the first book describing this new field of "stratigraphic analysis" of polar ice and devote his career to studying the various physical and chemical characteristics of deep cores. But in the summer of 1956 at Site 2, ice cores from the depths of a polar glacier were such a mystery, he recalled, that "nobody knew what they looked like, even." Putting the case more formally, Langway observed in his official report on the project that, while several exploratory expeditions had returned with valuable information on the upper few meters of old snow, "below the thin surface shell of the ice sheets lies an almost completely unexplored region about which little is directly known."

Success of the enterprise depended on a singular assumption: that the engineers could devise a mechanical drill capable of penetrating 1,000 feet or more into the ice with a hollow bit that would permit the extraction of core that was useful for careful scientific analysis.

However worthy the science, as a practical matter, how realistic was that goal? Researchers were about to discover that under extremes of pressure and temperature, ice, brittle and elastic in turn, behaves like no other substance on the planet. The job called for some creative engineering. As Army geologist G. Robert Lange described earlier efforts, "Results have been generally discouraging, with little or no usable core reported."

Summers in Greenland are cold and windy—miserable conditions for working with heavy equipment. The engineers dug themselves a big trench, 45 feet below the surface of snow, and covered it with a roof that the drilling mast punctuated like a spire. There was no escaping the intense cold, however, or the material contrariness of the ice. Drilling into deep polar ice was like working with a substance from another planet. Under such extremes of temperature and pressure, its physics are out of this world. One instant it turns to slush, the next it shatters like precious china. Boring an open hole would have been difficult enough, but extracting a well-preserved ice core demanded that roughneck drillers exercise the nearly impossible combination of brute power and extreme delicacy. During that first experimental summer in Greenland, even with all of the muscle and know-how of the U.S. Army at hand, and the nation's scientific and technical prestige at stake, there were times when an engineering solution seemed beyond reach.

Conventional drilling systems of the size and power required in Greenland and Antarctica had been designed for the petroleum industry to penetrate deep rock and sediment. The engineers were hoping that modifications to the heavyweight, industrial-strength tubular rotary drill would accommodate the special circumstances of extreme cold and the structural peculiarities of the ice. They were using what was known in the drilling business as a "Failing 1500," a rock drill graded to 1,500 feet depth, its pipe sectioned in 20-foot lengths. Langway would recall the Failing 1500 as "a terrible thing to manipulate" in the cold of northern Greenland. "You had to break pipe at 20 feet going down and coming up."

The drill employed compressed air as a rotating fluid to remove the ice cuttings, a circumstance that required several air compressors and a large heat exchanger to cool the compressed air. The chips of ice that melted from the frictional heat of the drill bit tended to refreeze and fuse together. Bits became stuck. Up and down it went, the entire string of pipe being dismantled with the retrieval of each section of ice. Much of the core was fractured or deformed by the process. Because the overlying burden put the ice under increasing pressure with depth, the deeper they drilled, the quicker the hole wanted to close. There were equipment failures and other unforeseen problems that are typical of elaborate operations in remote locales.

At the end of the summer of 1956, the drill had reached 305 meters—991 feet. Ice core was retrieved from about half that length overall. The deeper they drilled, the worse the ice core deteriorated. In his official report on the first season of effort, Lange accentuated the positive, promising that "although the quality of the core produced left a good deal to be desired, it appeared certain that improvements in quality and amount recovered could be assured by modification of equipment."

Although the drillers had proven that it was at least possible to penetrate a polar ice sheet to 1,000 feet and retrieve the core, these early results must have been privately disappointing to Henri Bader, who dreamed of opening the archive of climate history he believed was stored in the ice sheets. Only a fraction of the ice had been penetrated. While he could assure members of the U.S. National Committee that their commitment to deep-core drilling in Antarctica would be nominally successful, the results were not likely to be spectacular science. As he described the technical details, he told the science officials that a practical limit seemed to have been reached.

The experience of 1956 led to an overstatement of the problem. According to Bader, the Greenland trials had shown that the depth of 300 meters "is very close to the maximum depth for core recovery in any high polar glacier." Experiments by Langway at the SIPRE laboratories showed that at about 230 meters down the pressure from the overlying snow reaches a level whereby the pressure on the embedded bubbles of air gradually exceeds the tensile strength of the ice. Because the original drills used air to blow cuttings out of the ice core hole—suddenly exposing the deep ice to lower pressure—cracking began even before the core reached the surface. While later drillers did encounter a so-called brittle zone of ice susceptible to cracking between 700 and 1,300 meters depth, the use of a viscous liquid in the hole rather than air prevented wholesale fracturing of the core. (Ice from the "brittle zone" is stored for several months at the drill site to allow slow equalization of pressures before those sections of core are processed.)

In Greenland, what Langway's drillers intended as a single field season of drilling trials was extended to a second summer of experimentation. At the end of the 1957 summer season, in a new hole at Site 2, drillers reached 411 meters—1,336 feet. Although some practical limit seemed to have been reached, the drillers had achieved their goal of penetrating deeper than 1,000 feet into the ice sheet. Overall, this first attempt to drill deeply into a polar ice sheet was regarded as a qualified success. To a depth of 110 meters, they had extracted continuous undisturbed ice core, and down to 305 meters, 80 percent of the core was recovered. Below 360 meters, only two 5-meter lengths were usable core. Significantly, the quality of the ice core was much improved over 1956, and Chester Langway had more than enough polar ice core for his purposes.

In the Southern Hemisphere's summer of 1957-1958, American drillers, using techniques developed in Greenland and the same equipment, a modified Failing 1500 rock drill, reached a depth of 308 meters at Byrd Station, Antarctica. During the next austral summer, drillers at Little America V, on the Ross Ice Shelf, Antarctica, extracted core from a depth of 256 meters. While using the same modified conventional oil field drilling system, the engineers at this second Antarctic core-drilling site improved their results by filling the ice hole with a dense fluid to prevent the pressure of the surrounding ice from collapsing the hole.

Difficult as it had been, the IGY deep ice core project was a success. From the Northern and Southern Hemispheres, thousands of feet of polar ice core had been retrieved. Although the study of polar ice had not attracted widespread interest outside the U.S. Army cold regions laboratory, the National Science Foundation was sufficiently impressed with its potential to continue its financial support of polar ice research beyond the IGY.

Bader and Langway wanted to go deeper, of course, to plumb the ice sheets to their very depths. How many tens of thousands of years of Earth's climate history still were hidden far down in the ice? If they were ever going to know, they were going to have to solve some major engineering problems.

And here was a mystery worthy of the name: What if they were able to completely penetrate the ice cap? What would they find? Solid, liquid, or gas? Bader speculated that they would find, at the bed of the glacier, a volume of dry natural gas under tremendous pressure from the great weight of the ice. In that event, he wrote, the drilling crew "must be then ready to seal the hole to prevent development of a gusher." Russian scientist Igor Zotikov speculated that the Antarctic ice cap lay over great pockets of compressed air held down by its enormous weight. According to Zotikov, "The stores of energy of the compressed air which has accumulated during the time of the existence of the icecap is tremendous and could turn the giant turbines of a great electric power station for many thousands of years."

For a time, however, the distance down to bedrock must have seemed as far away as ever. Although the IGY projects had proven the concept of deep-core drilling, to the scientists and the engineers they had also proven that they were not very likely to get anywhere near bedrock in Greenland or Antarctica with a conventional rotary drilling system. Even if it had worked perfectly, transporting the heavy conventional drilling equipment with its 20-foot lengths of steel pipe to remote polar locales was practically impossible and intolerably expensive. If the research was going to survive long beyond the IGY, some innovative engineering solutions were going to have to be created.

In 1957, at the Army lab in Wilmette, Bader asked Lyle Hansen to begin work on a completely new drilling system designed specifically for extracting cores from deep polar ice. Since returning from the Taku Glacier expedition in the summer of 1950, Bader had been thinking about a drill that would take advantage of the physical particularities of ice—a *thermal* drill. With a souped-up electric transformer, they would try to melt their way through the ice cap. The electrically heated drill would be winched up and down the hole on a cable, doing away with the oil rigs' heavy and cumbersome lengths of continuous steel pipe. The problem of the collapsing borehole would be solved by filling it with a fluid more dense than the ice and with a lower freezing point—a viscous cocktail of Arctic-grade diesel fuel and trichloroethylene, a heavy toxic solvent. A vacuum pump would draw meltwater through heated tubes up into a tank above the core barrel.

In the summer of 1961, Chet Langway and the engineers took their new thermal drilling system back to northern Greenland and set up shop in the relative luxury of a curious experimental installation near Thule Air Force Base. Camp Century, the "City Under the Ice," was a secret Cold War installation that the Army had begun developing in 1959. Out on the ice of northern Greenland, 100 miles from Thule, a large and very elaborate military experiment was under way. With accommodations for some 200 men, enormous snow-covered trenches were laid out with dormitories, shops, theaters, clubs, and other amenities, including a well-equipped hospital and a well-stocked library, all powered by a little nuclear plant embedded in the ice. If the Soviet Union launched nuclear-tipped missiles at the United States, the thinking went, a counterattack could be mounted from this shelter under the ice. For such a critical Cold War mission, no expense was spared. The ice-coring project, expensive and difficult by standards of civilian science, must have seemed inconsequential at the time alongside grandiose military projects.

However, Camp Century was not long for this world. Movement within the ice sheet was deforming the tunnels, a circumstance that became increasingly troublesome and expensive. At one point, a team of about 50 men was devoted exclusively to keeping the tunnels open. More expense and trouble lay ahead. Under the terms of a treaty banning nuclear activity in Greenland and Antarctica, the nuclear plant was going to have to be removed. Although an alternate diesel-powered plant was available, the rapid closure of the tunnels was causing alignment problems in the power system. In 1966, the entire enterprise was abandoned, the little nuclear plant removed, and apparently just about everything else left to the crushing flow of the ice sheet.

About all that survives of Camp Century today is the name it gave to the ice core completed that last summer and the scientific history it made. By the summer of 1964, after trying three fluid-filled holes in the same trench at Camp Century, the crew had reached a maximum depth of 535 meters. The quality of the ice core was fairly good, although the thermal shock created stress in the core that caused additional fractures. By the third summer in northwest Greenland, it was clear to the crew that the thermal drill was not going to be the solution to the deep-core polar ice drilling problem that Bader and the Army engineers had hoped it would be.

A breakthrough came with a new oil well drill invented by Armais Arutunoff, who ran a little pump company in Bartlesville, Oklahoma. Suspended from a power cable, the electric motor and drive system was submersible in fluid. In 1965, the Army engineers took a reconditioned "Electrodrill" to Camp Century, inserted the diamond-cutting bit into the 535-meter hole, and quickly cored to a new record depth of 1,002 meters. During the next Northern Hemisphere field season, on July 4, 1966, the Army crew, using the Electrodrill, reached 1,387 meters—4,509 feet—and hit bedrock.

The new drill was then airlifted from Greenland to Byrd Station, Antarctica, where it reached the bottom of the Antarctic ice sheet at 2,164 meters in January 1968. During the following drilling season, however, when it was lowered back into the hole in an attempt to recover some material below the ice, the only drill capable of penetrating polar ice to bedrock became jammed and irretrievably lost. The United States would go for 20 years without a drill capable of producing a surface-to-bedrock core comparable to those garnered at Camp Century or Byrd Station, a circumstance that could have been a serious blow to the progress of polar ice analysis and the discovery of abrupt climate change were it not for the Danes.

At the University of Copenhagen, a team led by Niels Gundestrup designed and developed a deep-core drill system that finally would conquer the ice sheets of Greenland and Antarctica. Beginning in 1978, the three partners of the Greenland Ice Sheet

Program (GISP)—the United States, Denmark, and Switzerland—used the Danish drill in southern Greenland at Dye 3, the U.S. radar station that was part of the Distant Early Warning network arrayed across the Arctic Circle from northwestern Alaska to Iceland. In the summer of 1981, the GISP team reached bedrock at 2,037 meters—6,620 feet—a mile and a quarter from the surface. "The core is of excellent quality," Gundestrup reported, "and no part of the core is known to be missing."

In the 1990s, when the European nations of the Greenland Ice Core Project (GRIP) penetrated the ice sheet from the high-elevation Summit all the way to the bottom at 3,028.8 meters, a distance of nearly two miles, they again turned to the Danish ISTUK drill. Twenty miles away, the U.S. drillers of the Greenland Ice Sheet Program-2 (GISP2) reached bedrock a year later with a new drill developed at the University of Alaska, Fairbanks.

4

ROMANCING THE ICE

Inventing the science of polar ice analysis

In the 1950s, in the Wilmette, Illinois, laboratory of the Army's Snow, Ice and Permafrost Research Establishment, young Chester C. Langway Jr. experienced a rare and delicious moment for a scientist. Before him was a material that no other researcher had ever seen—ice from deep in the interior of a high polar glacier. What mysteries did it contain? Just as the roughneck drillers at Site 2 Greenland were out to prove that it was possible to drill deeply into polar ice, Langway was out to prove there were ways to analyze the ice cores that would yield information of scientific value. In the beginning, it was not so much a climate investigation as it was what researchers call a "proof-of-concept" project.

With Henri Bader looking over his shoulder now and then, it was Langway's assignment to tease from the ice information about its progressive age, to prove Bader's hypothesis that the cores contained a unique record of climates past and that the record could be read and understood. Everything depended on a basic idea that at this moment was unproved. It was not enough that the ice contained information about climates past. It had to present that information sequentially. If the past was to be reconstructed, the order of events had to be retained. In one form or another, the annual layering that was so evident and promising near the surface had to survive the transformation of snow into gritty granular material called firn and then into impermeable glacial ice. More than that, at depths where it could prove most interesting, it had to survive the unknown dynamics of thousands of years of ever-increasing pressure from an accumulating overhead load.

Nothing in his studies at Boston University had quite prepared Langway for this challenge. In the vernacular, his mission was to subject deep polar ice to what geologists call stratigraphic analysis, somewhat like studying the sediments in the bottom of lakes. The basic tools of geology and mineralogy were at hand, of course, but this was no pick-and-shovel job. What would the deep ice reveal? Langway could not know and Bader could not tell him what to expect. In the beginning, for all of the hope and promise and scientific salesmanship, and for all of the expense and the slow, slogging labor of drilling in bitter cold, it still was not entirely certain that deep polar ice would reveal much of anything.

The second summer of drilling in the far north of Greenland at Site 2 finally had produced, by August 1957, enough undamaged ice core from as deep as 411 meters to test the concept. Langway brought back 160 meters of core from various depths. Most of it was from near the surface, but he had enough usable deep core to settle the big question about the value—or folly—of drilling into polar ice sheets.

Ernst Sorge in 1930 and Bader himself in a field test in 1954 had seen enough old snow near the surface of the Greenland ice sheet to surmise that the annual layers were intact. Near the surface, at least, the archive was coherent—one year's accumulation could be distinguished from another. Below the surface, however, as the overhead load increases pressure with depth, snow is transformed into firn and then into impermeable, high-density glacier ice. Whether the annual layers survived this metamorphosis was an entirely open question.

It didn't take Langway long to confirm that the ice taken above 100 meters yielded to fairly straightforward methods of investigation. In ice cores taken from temperate glaciers in lower latitudes, where summer temperatures rise above freezing, researchers could observe unmistakable signs of dense summer melt and annual layers of windblown dust and pollen. But these convenient features are absent in glacial ice formed in regions of polar cold, where summer temperatures never rise above freezing. In the "dry-snow zone" of perpetually freezing polar cold, physical characteristics that mark annual layers are more subtle, produced by different processes.

Langway observed that in the dry zone the denser layer represents the winter snowfall, because it is colder, finer-grained, and more densely wind-packed. In the upper 100 meters of core, he was able to identify these regular density variations through the diffuse illumination of a light table, although an interesting reversal took place. In the back-lit effect of the light table, the lighter, whiter-appearing summer layer actually showed up darker than the bluish, more dense winter layer. Langway had the insight to recognize that the light was being scattered by the surfaces of air bubbles embedded in the ice, and the more porous summer layer was allowing less light to be transmitted through the core.

Langway identified the exact density at which the firn transformed into glacial ice and located this "zone of zero permeability" at 71 meters below the surface in the Site 2 core. But could he read the annual layers through this critical zone of transformation? Langway reported that he could, but only because he had continuous core to examine. He wrote: "A major factor in accomplishing the stratigraphic interpretations was the ability to trace continuously the structural features as a function of depth as the features and the surrounding material gradually metamorphosed from the undisturbed surface snow through the firn and into the high density glacier ice." Above 100 meters, the archive was intact and discernible. Langway counted annual layers reaching back 174 years.

For what he called "deep ice" below 100 meters, Langway already knew that "the classical stratigraphic approach" would fail him. He couldn't tell exactly where it would happen because he didn't have continuous ice core samples from below 100 meters, but he felt certain that structural features vanished and density variations were lost somewhere between 100 and 200 meters down. In fact, there was some question whether deep ice retained any features that could be related to original seasonal variations at all. New techniques would be required.

Laboriously, Langway studied the chemical composition of the ice. Here he found himself looking for impurities in one of the purest substances on Earth. "Deep polar ice is purer than triple-distilled lab water," he would recall. Measuring values in parts per million and parts per billion in small wedge-shaped samples of ice core was stretching the limits of mid-twentieth-century chemical technology.

He examined the microscopic structure and orientation of the ice crystals and the air bubbles they enclosed. Although this work did not advance his search, these clues to the age of the ice with depth, the air bubbles—capsules of ancient atmosphere—would prove to be a boon to later researchers.

Most importantly, Langway turned to a technology that was just coming into its own as a peacetime tool in the earth sciences: atomic isotope analysis.

Radiocarbon dating of different segments of the ice core would have been attractive were it not practically impossible. While the method was proving itself in calculating the age of ancient, carbon-rich organic matter such as wood, the state of the art at the time would have required a deep-core ice sample weighing a ton.

What interested Langway most was the ability of instruments known as mass spectrometers to measure accurately the proportion of the different stable isotopes of oxygen. Like a prism that separates the colors of light according to their different wavelengths, a mass spectrometer passes atoms through a magnetic field and spreads out their different isotopes according to their different atomic weights. Separately, two eminent researchers—geophysicist Willi Dansgaard at the University of Copenhagen in Denmark and physicist Samuel Epstein at the California Institute of Technology—were suggesting that the proportion of the rare heavy isotope ^{18}O to the common form ^{16}O in snow and ice is related to the temperature of its precipitation in high latitudes. A relatively warm temperature produces snow or rain of a relatively higher concentration of the heavier isotope. The differences are small—measured in parts per thousand—but a mass spectrometer can accurately detect them. The method had been tested in firn near the surface of Greenland's ice sheet and had revealed a swing in isotopic composition between summer and winter snowfalls. Whether a record of these seasonal swings survived in deep ice was an open question.

Bader wanted his ice core specialist to accompany the drilling operation to the IGY project in Antarctica, but Langway, having spent two summers in Greenland, begged off. Beginning in the fall of 1957, Langway devoted hundreds of hours at the Wilmette lab to the painstaking preparation of small, pie-shaped wedges of ice core for testing by Epstein's mass spectrometer. Langway prepared 439 samples in all and sent them off to Epstein.

Epstein's work at Caltech was proceeding slowly, and competition for use of the mass spectrometer was great. Langway started to wonder about the future of his project. In 1960, Bader left the Army Corps' lab for a professorship at the University of Miami. Where was Langway going to find the scientific help he needed to pursue his polar ice investigations? Three years passed before the data from Epstein's laboratory arrived.

Europeans always were more interested than Americans in ice. In 1962, at a meeting of the International Glaciological Society in Obergurgl, Austria, Langway was introduced to Hans Oeschger, a Swiss geochemist. Oeschger invited Langway to visit his laboratory at the University of Bern. Buried under 30 meters of earth and enclosed in tons of lead, the facility was designed to detect low-level concentrations of radioactive isotopes in nature. The two men discussed the possibility of using carbon-14, as well as radioactive isotopes of other elements such as beryllium and tritium, to assign ages to different depths of polar ice cores. Mostly, they just got to know each other and realized that they shared an enthusiasm for the study of ancient polar ice. Oeschger was a brilliant scientist and a quietly charming man. The collaboration that was formed by a handshake that day would hasten and enrich the study of climates past. Oeschger would lead the way to the discovery that bubbles of air trapped in the deep-core ice could be measured for their content of such greenhouse gases as carbon dioxide and methane—that they are direct "samples of the ancient atmosphere."

When the stable isotope analyses finally came from Caltech, the results were striking. In a plot of the isotope ratios, a strong seasonal variation was obvious, Langway reported, "with the summer layer richer in ^{18}O and the winter layers depleted in ^{18}O." Measuring the length of the waves during two consecutive seasons, Langway produced estimates of relative snow accumulation from one year to the next.

Langway could report with unqualified success that the new oxygen isotope method confirmed that "original seasonal variations established when the snow was first deposited are still preserved at depth." The archive was intact, deep in the ice, far back in time, an annual calendar legible down to the last bit of core at 411 meters, or 1,336 feet, which Langway estimated fell as snow in the tenth century, AD 934—coincidentally, about the time the Vikings were colonizing Greenland. He could only imagine what analyses of core taken from ice near the bedrock might reveal. Other U.S. researchers were slow to appreciate these findings, but as far as Langway was concerned, "the value of investigating deep ice cores is almost limitless."

When the drillers reached bedrock at Camp Century, Greenland, in 1966, a congratulatory letter arrived on Chet Langway's desk. The Army's facility had changed its name to the Cold Regions Research and Engineering Laboratory and moved from Wilmette, Illinois, to Hanover, New Hampshire, near the campus of Dartmouth University. The well-wisher was Willi Dansgaard. Langway knew Dansgaard only by reputation as one of the leading authorities in the field of stable isotope analysis. Dansgaard congratulated Langway on the technical feat of reaching bedrock at Camp Century and said he would be interested in conducting isotopic analyses of the ice core that had been extracted. He accepted an invitation to visit Langway's lab in Hanover.

When he arrived at the lab, Dansgaard told Langway that he had a contract with the International Atomic Energy Agency to study concentrations of tritium in water samples from around the world and that his lab was equipped like no other laboratory to analyze large numbers of stable isotope samples quickly. This was music to the American's ears. After three days, the two shook hands and agreed to work together on the Camp Century ice.

This arrangement between the American Langway, the Swiss Oeschger, and the Dane Dansgaard would prove to be one of the most fruitful international collaborations in twentieth-century science. The scientific talents, the technical capabilities, and the personal enthusiasm were in place for the systematic investigation of an archive of past climate that is unique in depth and detail. In their hands were the remnants of snow that fell before the last ice age. The antique window looking far out over the deep and varied landscape of Earth's past climates, first jarred loose by Ernst Sorge at Eismitte in the winter of 1930, was about to be flung open. Abrupt climate change was about to be discovered, if not entirely understood.

Dansgaard's analyses of the Camp Century ice core's oxygen isotope values in the late 1960s produced the first continuous record of Earth's climate going back more than 100,000 years. At the University of Copenhagen's mass spectrometry lab, scientists working around the clock had tested nearly 1,600 samples at 218 locations along 4,518 feet of ice. No other record on land or sea could match the Greenland ice sheet for detail or continuous timescale. Not even Antarctica would produce such a high-fidelity record.

The Camp Century core presented climate scientists with the first continuous profile of annual snow accumulation entirely through the last ice age, the Wisconsin, and into the pre-Wisconsin "interglacial" warm period known as the Eemian. When the results were first published in the American journal *Science* in 1969, even the wording of the title, "One Thousand Centuries of Climate

Record from Camp Century on the Greenland Ice Sheet," served notice to researchers who were accustomed to the geological timescale that they could expect a whole new level of detail from deep-core polar ice.

In the long upper section of core representing the 10,000 years or so since the last ice age, variations in the different values of heavy oxygen and light oxygen depicted the Younger Dryas and other lesser episodes of abrupt change, and these benchmarks helped calibrate the data with other climate records such as European pollens and lakebed and seafloor sediments.

In the lower segments, which are composed of snow that fell throughout the Wisconsin ice age, the core revealed a pattern that was nothing like the gradually sloping curves that most geologists would have expected. It was a staggering surprise, difficult to accept, a picture that would confound climate science for years. If the Camp Century core was an accurate archive of climates past, the prevailing view of ice ages as epochs of slumbering stability and slowly evolving change was not even half-right. "It showed all of these perturbations," Langway recalled. "At that time, it wasn't clear. Nobody knew what they meant."

The evidence of abrupt change is almost diabolically misleading. It depicts a climate shifting between different temperature and precipitation regimes at a pace that is totally unexpected. It has the raw and chaotic appearance of unrefined scientific data, which researchers know to be often messy and full of false starts and reversals, apparently meaningless noise. Scientists are trained to look for patterns in such disorder, to carefully select the signals that seem to bear some relation to one another, and so to offer hope of revealing a new clue about the orderly behavior of the world. At the end of the day, order is what science is about, and the ability to find order where others see only chaos is widely regarded as the mark of a great scientist. The idea that *the chaos is the signal* flies in the face of this tradition. It is just about the last thought that comes to mind—the "last survivor" of a process that has eliminated all of the more reasonable explanations of events. To the eye of an experienced researcher, at first flush the evidence of abrupt climate change naturally looks like data that are in need of more work.

In that first October 1969 article in *Science* and in a more thorough treatment of the oxygen isotope analyses of the Camp Century core presented at a December symposium the same year at Yale University, Dansgaard and his collaborators presented evidence of order where they could find it, tentatively observing periods of 120, 940, and 13,000 years and suggesting they were caused by variations in sunlight striking Earth. In their search for regularly recurring periods of warming and cooling, for "systematic oscillations," for the causes of climate changes on the large scale of geological time, the scientists were looking over the top of abrupt change. The telltale raggedness in the data was smoothed out by various mathematical filtering techniques. Instances of abrupt change that did not fit a pattern were characterized in the Yale paper as "accidental deviations that do not recur regularly," but Dansgaard was not going to ignore them.

In another brief article in 1972, Dansgaard took a closer look at the most dramatic swings of isotope variations from another angle: "Speculations About the Next Glaciation." Here was evidence that nobody had seen before. The beginning of the last ice age had come with a "rapid" drop in heavy oxygen isotope values, and some 15,000 years earlier an even more dramatic plunge to a cold climate had taken place. "Apparently, within 100 [years] the climate changed from warmer than today into full glacial severity," he wrote. Given the smoothing effect of molecular processes in the ice over time, the change may have come "almost instantaneously," and recovery from "this catastrophic event" may have taken 1,000 years.

The only other deep polar ice core, drilled at Byrd Station in Antarctica, did not reach as far back in time with continuous data. Although the deepest five meters of silty ice representing earlier ages contained several shifts between warm and cold climates, the Antarctic core did not display the kind of "violent oscillations" in oxygen isotope ratios that Dansgaard found in the Camp Century ice. And while the oldest Antarctic ice eventually would prove to be far older than Greenland's, the Greenland team noted "serious disadvantages" in the southern ice sheet. Most importantly, it is discontinuous: Whole years can go by without any new snow accumulating at such key stations as Byrd and Vostok. The fine detail of abrupt climate change shows up in the annually layered Greenland ice sheet like nowhere else in the world.

"The reason for the sudden changes is unknown," Dansgaard wrote. He ruminated about causes and effects. "Could ice surges from the Antarctic continent be responsible for an immediate and extreme cooling of the opposite Pole, in spite of the smoothing effect of the relatively slow coupling via the oceans?" The pattern of changes in the Byrd core suggested one possibility. Could it be that intense volcanic activity for some years "contaminated the stratosphere sufficiently to prevent most of the solar radiation from reaching the surface of the Earth?"

Why did one plunge in temperature usher in an ice age lasting tens of thousands of years whereas another did not? The big dips in isotope values seemed to be consistent with changes in Northern Hemisphere sunlight intensity caused by changes in Earth's orbit, although the dramatic climate changes seemed out of proportion with the subtle astronomical alterations. What mechanism would so drastically amplify such a signal?

Dansgaard asked: Were the sudden decreases in isotope values triggered by low incoming solar radiation, or *insolation,* in the Northern Hemisphere, where extensive areas of land can be covered by ice? "If so, the conditions for a catastrophic event are present today." He noted that the orbital cycle was moving into another period of low sunlight intensity in the north. "Or, are we faced with more or less accidental events, such as ice surges or intense volcanic activity, that trigger a full glaciation, if the insolation conditions favor such development? Is man's present activity equivalent to such [an] accidental event?"

The scientific rewards from the Camp Century ice core profile established not only the value to climate research of ice coring but more particularly the value of the Greenland ice sheet as a high-resolution climate archive. For sheer detail, no other place in the sea or on the land had produced such a picture of the past. The three researchers could only imagine what a finely resolved set of images would emerge from the profile of an ice core drilled in a Greenland locale chosen specifically for its scientific attributes—for its depth, its accumulation, its temperature, and the stability of its ice.

Greenland's value to climate science came as no surprise to the Scandinavians, of course, especially the Dane Dansgaard, who always chafed at the prevailing American attitude that Greenland was just an empty, featureless landscape, a big white blank on the map, of interest only as a Cold War military lookout base. With no particular affinity for the place, most American scientists saw Greenland the way their National Academy of Sciences had defined its role during the International Geophysical Year (1957-1958), as a logistically convenient proving ground for more scientifically important expeditions to Antarctica.

In the early 1970s, Langway, Dansgaard, and Oeschger embarked on an ambitious, long-term undertaking that was intended to plumb the great ice sheet from top to bottom. The three researchers formalized their collaboration with their national science funding agencies in the United States, Denmark, and Switzerland, respectively, and launched the Greenland Ice Sheet Program. Their plan was to survey the ice sheet as nobody had done before—to gauge its depth, of course, but more than that, to define its flow characteristics and the shape of the underlying bedrock. Like treasure hunters, they would sample the ice at a variety of sites and locate the best places to extract the longest possible continuous profile of climate history. They would drill three cores all the way to bedrock and then, in the field and their laboratories, explore all of the telltale chemical and physical features more thoroughly and carefully than ever, and tease from the ice every bit of data it had to give them.

A comprehensive aerial survey employing new radio-echo soundings, completed in 1974 by a Danish crew, produced the first full three-dimensional profile of the entire ice sheet. In 1975,

Langway moved from the Army laboratory in New Hampshire to the State University of New York at Buffalo, accepting an offer of tenure, a department chairmanship, and a new laboratory. From 1971 to 1978, the team spent every summer field season on the ice, experimenting with U.S., Swiss, and Danish drills, and sampled cores at 11 different locations, probing for the sweet spots, the places where the ice was likely to reveal the most about the history of the climates that had deposited its layers. They drilled shallow holes from one end of Greenland to the other, from the northern edge of the glacier high up near the pole at Hans Tausen, at 82°30′ North, all the way down to South Dome at 63°33′ North, a distance of 1,200 miles.

In his report at the 1969 Yale symposium, Dansgaard argued that the next core, the second core to bedrock, should be drilled in southern Greenland to compare it with Camp Century's far north polar conditions at 77°10′ North. Like Europe and eastern North America, southern Greenland is more directly influenced by the Atlantic Ocean than by the Arctic, Dansgaard observed, and such a profile "may contain more direct information about the conditions that led to buildup and extinction of the large Scandinavian and Laurentide ice sheets."

As time passed, however, increasing financial constraints forced on the GISP researchers the unhappy conclusion that they were going to be able to drill not three deep cores, as planned, but only one new core all the way to bedrock. "This limitation made it imperative that the drill site be chosen in the scientifically most favorable area to achieve the program objectives," they reported. Everything pointed to a central Greenland site high up on the ice sheet. In 1975, Dansgaard wrote: "Nowhere else in the world is it possible to find a better combination of reasonably high accumulation rate (which ensures continuity of record), simple ice flow pattern (which facilitates the calculation of the timescale), high ice thickness (which offers a detailed record, even at great depths) and meteorologically significant location (close to the main track of North Atlantic cyclones)." In 1972, the team had used a U.S. thermal drill to extract a core from a depth of 405 meters—1,316 feet—at a central Greenland site called Crête. This remote site was at the crest of the ice cap, 10,405 feet above sea level. Limited to one deep core, the team was hoping to begin drilling to bedrock at Crête in 1977.

Unfortunately, scientific imperatives were overcome by what at the time was seen as a more practical, less risky line of thought. The U.S. Army Cold Regions Research Laboratory had abandoned the ambitious Camp Century experiment, leaving just about everything but the nuclear power generator to be reclaimed by the flowing ice, and had bowed out of its role as logistical and transportation supplier. Now the entire expense of the expedition would have to be borne by civilian science funding agencies. Air transportation that used to be free was now costing $6,000 an hour. The researchers were going into the field with a new Danish ice core drilling rig that had not been tested at great depth, and as the price of operations climbed, the science agency bureaucrats were losing their appetites for risk. Duwayne M. Anderson, director of Polar Programs for the National Science Foundation, acknowledged that budgetary constraints dictated the choice of the Dye 3 site, where a U.S. radar installation already existed; although not an ideal location, it was of "sufficient interest" to justify the project.

Langway, Dansgaard, and Oeschger made the most of it. The new Danish electromechanical drill, battery powered and computer controlled, proved to be both reliable at great depth and fast. It reached 225 meters at the end of the field season in 1979 and 901 meters in 1980, and on August 10, 1981, bedrock was reached at 2,037 meters (6,620 feet), some 647 meters (2,123 feet) deeper than Camp Century. Handling the Dye 3 ice, the team set new standards for efficiency and analysis. While drilling was still under way, in a large "science trench" in the ice, they cut some 67,000 samples of core in a continuous sequence for oxygen isotope analysis. While still on the ice, the scientists tested the core for acidity, dust content, physical and mechanical properties, and chemical composition. Presenting their findings to a symposium in June 1982, less than a year after completing their field operations, the three scientists described Dye 3 as "the most completely recovered, thoroughly recorded, and comprehensively investigated ice core yet to reach bedrock."

Some—but only some—of the same abrupt changes first revealed by the Camp Century core had shown up in ice cores drilled on Devon Island, in Arctic Canada, across Baffin Bay from northern Greenland. Did the oscillations in Camp Century's oxygen isotope values really depict rapid climate changes in the Arctic? The question remained open for the better part of a decade, until the recovery of the core at Dye 3 gave Dansgaard, Osechger, and Langway the corroboration they needed.

"Most of the pronounced oscillations observed in the oxygen isotope profile from the deeper part of the Camp Century ice core were reproduced in the Dye 3 core, confirming their climatic significance and, at least, the regional validity of these data," they reported. Dansgaard took a close look at the evidence of rapid climate change in both cores—the jagged patterns of changing oxygen isotope ratios, designated δ, through the Wisconsin ice age—and found the pattern of correlation especially significant in view of the "completely different ice flow conditions" in two areas nearly 900 miles apart.

Looking for a cause for the sharp changes in isotope ratios, Dansgaard ruled out any local climate conditions in Baffin Bay or ice sheet instability, pointing instead to general climatic changes in the Arctic, "perhaps due to alternations between two different quasi-stationary modes" of the climate system.

Hans Oeschger observed how the temperature-related oxygen isotope variations in the ice and carbon dioxide concentrations in the trapped air bubbles switched back and forth in lock step with one another. He saw how the cores revealed the same patterns of abrupt change known as the Younger Dryas in the lake sediments and peat bogs of Europe. It was all part of this curiously "bistable climatic system." What could explain such abrupt changes?

In a separate paper in 1984, Oeschger and colleagues at the University of Bern, joined by Dansgaard and Langway, described the different timescales in the patterns that emerged along the oxygen isotope profile and proposed different causes for them. Although the idea that "changing seasonal solar irradiation due to variations of the earth's orbital parameters is the primary cause for the glaciation cycles" was gaining ground with new data, the corresponding changes are "relatively slow compared to the climate fluctuations revealed in the Greenland ice and in European lake sediments," they wrote. "These events, however, can be linked to oceanic changes, namely advances and retreats of cold polar water in the North Atlantic Ocean." The southward advance of polar water coincides with colder climate in Europe, they noted, and its retreat with a warmer Europe. "This strongly suggests that the continental climate shifts were an effect of changes of the surface conditions of the North Atlantic Ocean."

From out of the Greenland ice, a new picture was coming into focus. The landscape is not smooth. The grand swayings that appear to define the coming and going of the ice age are riveted with large staccato punctuations of abrupt change. It is not a matter of the pace of change picking up and then slowing down. All along the way, different drummers are pounding out different rhythms. Different mechanisms are at work.

The Camp Century core, the Dye 3 core, the lake sediments in Switzerland, the pollen profiles in Denmark and elsewhere in Scandinavia all told the same story. Suddenly Earth's climate began pulling out of the ice age about 14,700 years ago. Then, just as suddenly, after only about 2,000 years, it plunged back toward glacial conditions for a thousand years or more. And then, abruptly, climate conditions recovered and began a more gradual warming toward the relative stability of the past 10,000 years. The scientists wrote that each of the three climate changes is marked by an abrupt shift in oxygen isotope ratios, signifying drastic changes in central and northwestern Europe at those times.

Duwayne Anderson at the National Science Foundation congratulated the scientists and engineers of the Greenland Ice Sheet Program for "an effort that now can be described as brilliant" and for their papers that "provide a 'shower' of knowledge and information. . . ." Thirteen years of work on Camp Century and then

GISP had brought new methods of analysis, new technologies, and a whole new realm of data to a fascinating new branch of earth science, christened "paleoclimatology." A provocative new concept of climate change was born.

As a stand-alone science product, the Dye 3 core itself would not be seen as a singular contribution to the archive of Earth history. Its greatest value was in corroborating many of the details revealed first and more effectively at Camp Century, by a core that continues to challenge researchers into the twenty-first century.

In the end, nothing could entirely rescue the Dye 3 profile from the fact that this primary core to bedrock was drilled at a location on the ice sheet that failed the principal scientific criteria laid down by the researchers themselves. Given that they were getting only one core to bedrock, it was simply a bad place to drill. "Complicating the ice core analyses is the extremely complex ice flow trajectory over the hilly bedrock upstream of the recovery site," they wrote. "This required further field work in 1982 and 1983 in order to establish two of the fundamental time-series (i.e., annual snow accumulation and temperature change)."

Being so far south, the Dye 3 location was too warm. In the current climate, at least, summer temperatures often rise above the melting point of water, and atmospheric carbon dioxide readily dissolves in the meltwater, leaving spuriously high concentrations of the gas in the affected ice. The scientists lamented the fact that "this situation invalidates some of the Holocene record (past 10,000 years) of the Dye 3 core as a source of valid information on one of the most urgent problems in modern climatology, i.e., the interaction between climate and atmospheric CO_2, including the turnover of CO_2 in the terrestrial reservoirs." The scientists thought, but couldn't be sure, that the Dye 3 region probably was too cold for summer melting during the ice age.

Finally, even though the core retrieved was in better condition than Camp Century's, the Dry 3 profile did not reach as far back in time, at least not in continuous sequence. Missing from Dye 3 was ice deposited on southern Greenland before the beginning of the last ice age, during the last interglacial, when climate was more like the current Holocene regime. Climate scientists call this period, about 120,000 years ago, the Sangamon, or Eemian. Dansgaard reported that the mountainous bedrock upstream of Dye 3 had disturbed the layers in the deepest 87 meters of the core. "This makes it impossible to study a time period of great scientific interest: the Eem/Sangamon interglacial and its termination. Detailed studies of this event are important, because they might shed light on the significant problems of why, how, and when the present interglacial period will end."

5

"NINE AND SIXTY WAYS"

Theories of ice ages past and future

What causes a warm world of flowing water and verdant growth to become a cold world of dry winds raking arid landscape? This question has been the object of a long and continuing rumination, of more than a century of field investigation, theorizing, and debate. Along the way, a new history of Earth's climate has been written and a science of paleoclimatology invented. And yet, for all of the new techniques and powers of reasoning at their command, researchers cannot satisfactorily answer the basic question: Why do such large changes in climate periodically overtake the planet?

The difference between one world and the other is about as different as worlds can be. The difference is mile-thick ice burying 11 million square miles of continent—Asia, North America, and Europe sagging under its weight. The difference in the level of the oceans is 400 feet. The difference is a mere 9°F in mean global temperatures. What these parameters of modern climate science fail to describe, of course, is the difference that is most obvious and most germane to humankind—that one is a world with much more life in it than the other.

Like the great swayings of the ice ages themselves, theories about them have come and gone. Early in the century, the British meteorologist C. E. P. Brooks borrowed a line from Rudyard Kipling to describe the state of the art. "There are at least nine and sixty ways of constructing a theory of climatic change," he wrote, "and there is probably some truth in quite a number of them." U.S. Weather Bureau physicist William J. Humphreys lamented that "no one can form the slightest idea" about past and future ice ages—when they might begin, how cold they may be, or how long they may last. Some fine minds have visited the problem, although in the first half of the century, Brooks and Humphreys were among only a handful of researchers who were knowledgeable in the ways of the atmosphere. For most, climate change was not a serious issue, and ice ages were ancient, irrelevant history. Most people who called themselves climatologists were busy tabulating the statistics of weather for the benefit of local farmers and engineers, and most meteorologists were practicing a trade of weather prognostication founded on maps of pressure differentials and intuition.

In Brooks's day, climate research was a poor, young, backwater branch of geology in competition for students with the lucrative oil exploration business, a venture that takes a different interest in strata bearing signs of ancient vegetation. Without the data to test an idea or the technical wherewithal to pursue it, researchers proposed theories of climate that, for many years, were what scientists politely call *qualitative*—little more than hand-waving arguments and just about anybody's game. Carl A. Zapffe, a Baltimore metallurgist, mixed his physics with esoteric theology and the legend of the Lost Continent of Atlantis to bolster his "Submarine Vulcanism Theory." Periodically, volcanoes would erupt along the floor of the Atlantic and spew vast blankets of cooling dust into the atmosphere at the same time they warmed the sea, promoting evaporation, a boost in water vapor, and increased snowfall.

To finally solve such a large and enduring puzzle would be a mark of the greatest scientific distinction, worthy of a Nobel Prize. The hunt would be joined by astronomers, mathematicians, physicists, geologists, oceanographers, biologists, botanists, chemists, climatologists, meteorologists, and computer modelers. They would look everywhere for explanations—deep in the earth and far out in space. They would consider the multifaceted character of the climate system itself—how the oceans, the atmosphere, and the ice cover affect one another—and any number of external astronomical possibilities.

For all of the early uncertainties, most of the main theories of climate change were proposed remarkably soon after Louis Agassiz's famous "Discourse" in 1837, when in Neuchâtel, Switzerland, he unveiled his theory of the ice ages. As different types of data accumulated over the years, theories would shift in and out of favor among the researchers of the day. Much of the history of the thinking about ice ages has followed this pattern of the debunking of one venerable theory and the resurrection of another.

Among the first explanations were the geological ones. The nineteenth-century British geologist Charles Lyell and later the American James D. Dana pointed to mountain building and other changes in the heights of Earth's crust as possible explanations. One theory held that the poles have changed location over time, due to changes in the planet's tilt, exposing different flanks of Earth to warm and cold latitudes. In the 1920s, Alfred Wegener and his father-in-law, climatologist Vladimer Köppen, offered a variation on that theme. The Germans explained ice age evidence in terms of Wegener's theory of continental drift. Like the "wandering pole" theory, climates of the continents were altered as they drifted through different climate-controlling latitudes. At the U.S. Weather Bureau, Humphreys persuaded himself—although not very many of his colleagues—that the cooling effect of dust from volcanic eruptions was to blame.

Several theories were linked to processes and features internal to Earth's climate system. The role of the gas carbon dioxide in the atmosphere was first proposed as a cause of ice ages by Swedish researchers Svante Arrhenius and Nils Eckholm at the turn of the century. Pointing to the "greenhouse effect" of CO_2, which is transparent to the Sun's short-wave radiation but traps long-wave (i.e., heat) energy radiating back from Earth's surface, they argued that a loss of carbon dioxide could send the planet into an ice age and that the buildup of CO_2 could stave off such a catastrophe. A century later, of course, the close relationship between temperatures and greenhouse gases would become critical in the context of another climate puzzle.

Early in the century, meteorologist Brooks suggested a meteorological explanation—the idea that the formation of ice provoked a feedback mechanism that progressively cooled the planet by reflecting more and more sunlight from its surface. In 1956, Maurice Ewing and William Donn gave this idea another twist when they proposed that ice ages were triggered by rising Arctic temperatures. When ocean temperatures warmed to the point at which the Arctic became free of ice, a cascade of atmospheric effects—heightened evaporation and increased water vapor—led to a surge of snowfall over nearby lands, to the ice reflection feedback loop and the spawning of a new ice age.

Widely publicized at the time, Ewing and Donn's theory was the first expression of a concept that would become a common scenario in a globally warming world. Columbus Iselin, director of the Woods Hole Oceanographic Institution, where Ewing and Donn were doing their work, set the stage in a 1957 interview with *Christian Science Monitor* journalist Robert C. Cowen. Noting that the modern era was "literally flooding the atmosphere with carbon dioxide," Iselin asked, "Are we making a tropical epoch or are we, perhaps, starting another ice age? We don't know enough about the oceans and the weather yet to be sure which way the effect will go." Later scientists and journalists would sketch a similar scenario. Warming temperatures in the far north would not necessarily usher in an ice age, but could spread colder-than-normal temperatures over much of northern Europe.

In 1964, the New Zealand scientist Alex T. Wilson suggested that the Antarctic ice sheet occasionally reaches critical mass, collapses, and sends large sections surging into the surrounding ocean. Sea levels rise and sunlight bounces off the spreading sheets of floating ice, cooling the atmosphere. While researchers found no evidence of such an event in the sediments of the Southern Ocean, Wilson was among the first to recognize the ability of ice sheets to fail catastrophically, an idea that would become important years later in the search for mechanisms of abrupt climate change.

Among the most powerful and popular of the early theories were those that looked beyond Earth to astronomical circumstances. One proposal was from the realm of celestial mechanics. Certain geometrical features of Earth's orbit of the Sun could account for the big swayings, the long timescale cycles in the ice age data. More directly, another proposal suggested that changes were due to variations in the Sun's luminosity, suggesting a link between the appearance of sunspots and differences in the amount of heat reaching Earth.

At the beginning of the nineteenth century, even before Agassiz began to establish Earth's ice age past, the English astronomer William Herschel was speculating about solar variations, although the very idea encountered deep-seated resistance. Like their climate, what humans want of their life-giving star is stability. By the middle of the nineteenth century, astronomers at least offered regularity, establishing the cyclical nature of sunspots and identifying an 11-year period, a pattern that suggested a link to weather if not climate.

A century later a long, tedious, star-crossed campaign to find correlations between sunspot cycles and changes in weather was under way. Led by American astronomer Charles Greeley Abbott, 1920s and 1930s researchers identified weather-sunspot correlations just about everywhere they looked. For many years, Abbott insisted that he could detect variations in the Sun's luminosity and that, before long, these variations would lead to great improvements in weather prediction. Yet every attempt at prediction was a failure. As Theodore S. Feldman wrote in a recent essay on the subject, "These attempts got a well-deserved bad reputation."

For climate researchers, the search for links to variations in the Sun's output has been more respectably cautious, although only a little more rewarding. While the Greenland ice sheet record would prove to be too short to help resolve many puzzles about long-term ice age patterns, its clarity of detail promised valuable insights into changes on the order of 10 years, that is, almost the duration of the 11-year sunspot cycles. The Camp Century ice core team weighed in on the subject in 1970 with an article by Sigfus J. Johnsen, Willi Dansgaard, and Henrik B. Clausen of Denmark and Chester Langway of the U.S. Army's Cold Regions Research Laboratory. The stable isotope variations going back to the year AD 1200 revealed oscillations with periods of 2,400, 400, 181, and 78 years, they wrote, and all of the periods "seem to originate from changing conditions on the Sun." In a graph depicting their isotope results, the Greenland team boldly extended the cycle out into the future, drawing a curve that showed a cooling of global temperatures to the year 2000.

At Lamont-Doherty Geological Observatory in New York, Wallace S. Broecker looked at these Greenland results and even more ambitiously predicted a different result. "It is possible that we are on the brink of a several-decades-long period of rapid warming," Broecker wrote in 1975. The cooling of global temperatures that had been under way since 1940 was "one of a long series of similar natural climate fluctuations" seen in the Camp Century core that had been counteracting the warming effects of the increasing concentration of carbon dioxide. But the natural trend was about to "bottom out," he wrote. "Once this happens, the CO_2 effect will tend to become a significant factor and by the first decade of the next century we may experience global temperatures warmer than any in the last 1,000 years." Looking back over 25 years of warming, Broecker would acknowledge years later that he had "made a large leap of faith" in assuming that the Camp Century cycles of 80 and 180 years were global. Science historian Theodore S. Feldman described the 1975 prediction as "one of several cases where Broecker's scientific instincts were sounder than his evidence."

The short-term Camp Century oscillations, whatever their cause, did not materialize as global phenomena in other data, and the accumulation of other evidence did not back up their supposed link to variations in the Sun's brightness. Decades later, scientists were willing to assume that the minor variations in solar radiance that had been measured by orbiting satellites are among a variety of forces contributing to changes in a climate that already is unstable. As a mechanism that explains ice ages, however, the variability of the Sun had given way to a more powerful astronomical theory.

Since the mid-1970s, most researchers have satisfied themselves that the pattern of climate change at the geological timescale is best explained by one of the oldest theories. Its main elements are features of celestial geometry—the shape of Earth's orbit and the tilt in the planet's axis, that is, its angle of rotation relative to the Sun.

As Johann Kepler established in the seventeenth century, Earth's orbit of the Sun is slightly elliptical rather than circular. The Sun occupies a focus of the ellipse that is offset from the center, which means that during one time of the year the Earth is slightly closer to the Sun than it is at another. As Earth orbits the Sun annually, its elliptical orbit itself also rotates about the Sun, but on a much longer cycle. On another timescale, the orbit changes shape slightly, becoming more or less circular. Every 24 hours, of course, the Earth completes its rotation around its axis, which is currently tilted at a 23.5 degree angle in relation to its orbital plane. That angle, the feature that accounts for the different seasons, varies slightly, between 22 and 25 degrees, at one timescale. On yet another cycle of different duration, Earth's tilting axis wobbles, like a top spinning down. All of these geometric features and their different rates of change—complex in combination but mathematically predictable—alter the intensity of sunlight striking Earth.

The interplay of these properties of celestial geometry produces three periods of oscillation in the climate record—at about 22,000 years, 41,000 years, and 100,000 years—and forms the basis of the generally accepted explanation of the ice ages. This is what scientists refer to as the theory of orbital or astronomical forcing, or the Milankovitch theory. Agreement about it has been a long time coming, and it is still incomplete. As a theory of climate change, it is not very satisfying. One of the most important scientific papers on the subject describes these elements collectively not as the cause but as the "pacemaker" of the ice ages.

In 1842, just five years after Agassiz's "Discourse," Joseph Adhémar, a mathematics teacher in Paris, was first to suggest that the ice ages were related to Earth's orbit of the Sun. The changes of Earth's wobbly tilt through its elliptical orbit meant that the hemispheres alternately experience somewhat shorter summers and longer winters through a 22,000-year cycle. Focusing on this cycle, Adhémar surmised, mistakenly, that one hemisphere would get colder as the other warmed. He sketched an improbable scenario in which ice ages came and went with the waxing and waning of the Antarctic ice sheet. About all that survives of this first theory is the basic premise that astronomical circumstances are involved.

In the 1860s the Scotsman James Croll extended and strengthened the astronomical theory to include very slow changes in the elongated shape, or *eccentricity*, of Earth's elliptical orbit. Croll argued that ice ages are triggered during epochs when Earth's orbit is most elliptical, in its winter hemisphere, when it is farthest from the Sun. During these epochs, he theorized, ice ages occurred in alternating hemispheres every 11,000 years. Calculating these periods of high eccentricity of the orbit, Croll estimated that the last glacial epoch ended about 80,000 years ago. Although meteorologists questioned the power of relatively minor changes in the shape of Earth's orbit to provoke such dramatic climate changes, Croll identified two amplifying effects—the reflection of sunlight off growing ice sheets and the increasing strength of the equatorial trade winds. Geologists debated the theory for years, but support for it waned as evidence slowly accumulated that the last ice age had ended about 10,000 years ago, rather than the 80,000 years estimated by Croll. By the end of the nineteenth century, scientists had discarded the astronomical theory as an explanation of ice ages.

The theory was resurrected in the 1920s by the Serbian mathematician Milutin Milankovitch, who carefully developed a set of curves that measured the orbital effects on radiation at different latitudes over the past 650,000 years. His work gained widespread recognition when the German climatologist Vladimer Köppen and his son-in-law, Alfred Wegener, incorporated Milankovitch's calculations into the publication of their own research into the geological history of climate. To Köppen and Wegener, the Milankovitch diagrams seemed to fit the evidence, which German geographers had gathered from gravels left by alpine glaciers, that identified four major ice ages during the past 650,000 years. Also, rudimentary estimates that the last European ice sheets disappeared about 20,000 years ago seemed to agree with the Milankovitch curves. For 50 years the pattern accepted by geologists was one of long, warm periods interrupted by four relatively brief ice ages.

Powerful new technologies developed after World War II changed the character of the ice age debate, allowing geologists to more fully test the astronomical theory on their own terms. As Milankovitch himself had observed in 1941: "These causes—the changes in insolation [that is, solar radiation reaching Earth] brought about by the mutual perturbations of the planets—lie far beyond the vision of the descriptive natural sciences." To geologists, who were unfamiliar with the realms of celestial physics and mathematics, the postwar technical advances brought the theory "down to Earth," in a sense, by giving them new geological tools. Nuclear technology brought radiocarbon dating and stable isotope analysis. The Swedish geologist Börje Kullenberg designed a new piston-core device that could retrieve much longer ocean sediment cores. And the discovery of magnetic reversals in the geological record made it possible to correlate data around the globe.

Radiocarbon dating gave researchers their first relatively accurate age estimates going back 40,000 years. In 1957, Wallace S. Broecker had made the application of radiocarbon dating to climate questions the subject of his Ph.D. thesis. He would spend 10 years as director of Columbia University's radiocarbon laboratory at Lamont Geological Observatory and would play a key role in developing theories about both the slow roll of ice ages and abrupt climate change mechanisms.

As it happened, the first radiocarbon results provoked another sudden collapse in scientific support for what had become widely known as the Milankovich theory. In the mid-1950s, geologists discovered a peat layer in Farmdale, Illinois, that was dated at 25,000 years old. Here was a relic of a warm climate that came precisely at a time when Milankovich's calculations called for a solar radiation minimum. Similar discrepancies followed elsewhere in North America and Europe, and by the early 1960s the astronomical theory of the ice ages again was in eclipse.

Broecker did not give up on the concept, however. In 1965, in a conference presentation entitled "In Defense of the Astronomical Theory of Glaciation," Broecker argued for a new model of climate that made important changes to Milankovitch. The climate system has "two stable states, glacial and interglacial," Broecker proposed, and "rapid transitions between these states" are triggered when Earth is receiving maximum radiation from the Sun.

Developing ideas he had proposed in his Ph.D. thesis, Broecker was challenging two tenets of conventional thinking on the subject. The German model of four relatively brief ice ages widely separated by much longer warm periods had given rise to the idea that Earth's climate was in some way naturally inclined to be warm unless perturbed by some exceptional circumstances. And almost invariably, whatever their magnitude, changes could be depicted along a timeline of gently sloping curves. In the years ahead, these two ideas—of two stable climate states and rapid transitions between them—would become central to understanding abrupt change.

In 1968, Broecker helped revive interest in the Milankovitch theory with radioactive isotope tests he conducted on samples of ancient coral taken in Barbados by Robley K. Matthews. Terraces of coral reefs that stand far above the present sea level cover much of the Caribbean island, and Matthews had established that the different terraces were relics of eras of higher sea level. Applying his nuclear laboratory methods to a slow-decaying isotope of thorium, Broecker established that the different terraces of coral reef were formed 125,000, 105,000 and 82,000 years ago. According to the Milankovitch curve for the latitude of 45° North, these were times of maximum solar radiation intensity, when melted ice sheets had caused exceptionally high sea levels. Reporting these results, Broecker called the Milankovitch theory "the number-one contender in the climatic sweepstakes."

At the time, even more compelling information was coming from the sea itself. The technical improvements in the retrieval of ocean floor sediments had brought a whole new archive of ancient climate into the laboratories. These sediments, accumulating at a rate of a few centimeters per thousand years, were proving ideal for studying a problem on the timescale of ice ages. Although their laminations did not reveal the detail found in polar ice, especially Greenland ice, the seafloor sediments were much older, and the new technology allowed researchers to extract long cores that revealed large-scale climate changes going back millions of years.

Researchers used two techniques to tease climate changes out of the sediments. The cores are composed primarily of calcium carbonate skeletons of minute ocean creatures known as foraminifera, or forams, which settle to the seabed over the eons, and these tiny skeletons serve as proxies for changes taking place in the climate over time. Like the polar ice investigators, geologist Cesare Emiliani and geophysicist Nicholas Shackleton employed the new nuclear technique of mass spectrometry that would prove so valuable in Greenland. In the calcium carbonate, they measured the changing proportions of the stable isotopes of oxygen, believing that variations between the common ^{16}O atom and the rare and heavier ^{18}O atom revealed ocean temperatures at the time when the forams lived. Like the German Wolfgang Schott, who pioneered the field in the 1930s, geologist David B. Ericson at Lamont Geological Observatory sampled populations of the forams at different depths and built a temperature profile according to the assemblages of warm-water and cold-water species.

In 1966, Emiliani reported sediment results that he described as a record of ocean temperatures through the last 400,000 years. This research, begun in the mid-1950s, fundamentally changed the picture of ice ages held by two generations of geologists, although it was mired in controversy for years. Eventually, most scientists agreed with Shackleton and others that Emiliani's results reflected changes in global ice sheet volume as well as ocean temperatures. The critics pointed out that because the lighter ^{16}O isotope evaporates more easily from the ocean's surface, the buildup of great ice sheets caused the subsiding seas, whatever their temperatures, to be enriched with ^{18}O. As Willi Dansgaard pointed out, however, a profile of "paleoglaciation" was just as valuable as a profile of "paleotemperatures." For identifying ice ages, what could be better than a reliable record of the comings and goings of glaciers and ice sheets?

For a time the controversy over the meaning of the data tended to obscure Emiliani's more fundamental discovery. His sediment profiles identified many more ice age periods than the conventional four ages described by the German geographers at the end of the nineteenth century. It challenged the very basis of Köppen's and Wegener's support for Milutin Milankovitch in the 1920s. Although the Emiliani curves agreed with the Milankovitch diagrams, they turned the prevailing conception of past climate on its head. Rather than four ice ages, there were many. Rather than an equable climate occasionally interrupted by ice ages, the new pattern was the opposite. More frequent and longer-lasting ice ages were punctuated by much shorter warm periods.

In 1968, the Czechoslovakian geologist George Kukla demolished the old order. Kukla had studied layers exposed at a brick quarry and found ice age layers of windblown silt more frequent and numerous than would be expected by the German model. In fact, he found a frequency of 100,000 years in the ice age cycle. Using the new techniques, Kukla returned to the alpine terraces the Germans had used to construct their ice age profile and found that the gravels contained material of very different ages than they had supposed.

In 1970, Wally Broecker and his Lamont colleague Jan van Donk reported results of isotopic dating of a core taken from the Caribbean that agreed with Emiliani's multiple ice age data as well as the Milankovitch curves. They identified a saw-toothed pattern in the profile, representing long, slow descents into ice age cold that ended in "rapid terminations." The dominant cycle lasted about 100,000 years.

Another important breakthrough came in 1973, when Nicholas Shackleton presented a climate profile he had meticulously developed from one of the best sediment cores ever extracted from the seabed. Taken in 1971 by the Lamont research vessel *Vema* from shallow water in the western equatorial Pacific, the core reached back more than a million years, beyond the time of the last magnetic reversal 700,000 years ago. A new isotope analysis method using radioactive potassium detected evidence of the event in the core. This critical feature allowed Shackleton to calibrate climate events along the time profile of the core, establishing the first accurate chronology of the last 1.5 million years of climate. This "Rosetta Stone" of ice age research also showed a dominant 100,000-year cycle.

In 1976, Shackleton, John Imbrie, and James D. Hays put it all together in the classic paper on the subject in the journal *Science*. Their study of two cores from the South Atlantic showed Milankovitch curves at 23,000, 41,000, and 100,000 years that coincide with changes in the geometry of Earth's orbit. The scientists concluded that "changes in earth's orbital geometry are the fundamental cause" of the succession of ice ages, although they frankly acknowledged that they could not explain just how these changes affected climate.

The study confirmed a set of particularly puzzling facts. The cycles that imposed the strongest differences in the strength of sunlight reaching the planet, the 23,000- and 41,000-year cycles, together accounted for about 35 percent of the climate variation over the past 500,000 years, they estimated, and probably were strong enough to influence climate directly. On the other hand, the dominant 100,000-year cycle which they associated with fully 50 percent of past climate variation was too weak to cause these changes directly. In the vernacular, the response was *nonlinear*, that is, the rules of traditional physics don't apply. One unit of push does not necessarily produce one unit of shove; the system is chaotic and huge changes are possible from minor variations. Something about the climate system caused it to respond all out of proportion to the minor difference in solar radiation that results from slight changes in the elliptical shape of Earth's orbit. Or something else was causing the 100,000-year cycle.

Scientists like to use the word *elegant* to describe a theory that has the power to reduce complex circumstances to simple explanations—in a single stroke, to bring order from chaos. Plate tectonics is thought of as elegant, for example, as is Darwin's theory of evolution by natural selection. As it happens, however, on the basic question of the causes of ice ages, even when beautifully predictable properties of celestial mechanics are invoked, Earth's climate evidently is not congenial to elegant explanations.

Still, there are lingering doubts, especially about the cause of the 100,000-year cycle, and major mysteries remain unsolved. If anything, this best explanation for ice ages raises the mystery to a new level of complexity. There would be no *Eureka!* moment for the ice age problem, no occasion when someone would jump up with a solution that would transform the science. Researchers instead were left with the unsatisfying prospect of plowing through the data to identify any number of mechanisms that could amplify this insignificantly weak radiation signal so powerfully from afar. *Nonlinear* meant that the future, for the time being at least, was just about anybody's guess.

The Milankovitch orbital timetable led Shackleton, Imbrie, and Hays to predict the onset of an ice age; in their words, "the long-term trend over the next several thousand years is toward extensive Northern Hemisphere glaciation." They warned, however, that forecasts based on the astronomical theory came with two critical qualifications. "First, they apply only to the natural component of future climatic trends—and not to such anthropogenic [or human-caused] effects as those due to the burning of fossil fuels. Second, they describe only the long-term trends, because they are linked to orbital variations with periods of 20,000 years or longer."

By the early 1970s, the future of the climate was becoming an issue of increasingly widespread concern. A powerful El Niño in 1972 so wrecked the grain harvest in the Soviet Union and elsewhere that modern leaders for the first time began to wonder about the security of the world's food supply. A distinct cooling trend had been under way since 1940, and magazine and newspaper articles were raising the specter of a new ice age in the minds of the general public. Moreover, recognition of a climate history dominated by successive ice ages was giving rise to a different picture of the future than most everyone had assumed. The concept of abrupt change had appeared on the horizon, like an uninvited guest. Dansgaard had reported early data from the Camp Century ice core. And two contending ice age theories—Ewing and Donn's Arctic ice scenario and Wilson's Antarctic ice surge—suggested that changes could be both rapid and devastating.

In January 1972, leading climate scientists gathered at Brown University in Providence, Rhode Island, to pool their thinking about the future of the current 10,000-year-old warm "interglacial" climate episode that had cradled the rise of civilization and agriculture. The conference—The Present Interglacial: How and When Will It End?—attracted most of the major climate scientists. Among the contributions was "Speculations About the Next Glaciation," a brief paper by Willi Dansgaard and both American and Danish members of the Greenland ice core team.

Like scientists before and since, the researchers who threw open the Camp Century window and first saw the fine detail in the landscape of climates past had to set their sights over the tops of the canyons and ridges of abrupt change in search of the more distant ranges and divides left by the advances and retreats of glaciers over geological time. Even then, they could not avoid the most striking feature that the variations in the oxygen isotope ratio revealed in the Camp Century ice—a pace of change that had never been seen before.

The Dansgaard team noted that an event they estimated to have taken place 89,500 years in the past had plunged the climate "from warmer than today into full glacial severity" within just a century or even less. In fact, the drop in the oxygen isotope ratio, and thus in temperatures, might have occurred "almost instantaneously." The curve in the climate profile, they wrote, "suggests that it took 1000 [years] to recover from this catastrophic event." Like no other record before it, the Camp Century ice profile was recording climate events on a timescale of centuries and decades, the units of human history. "Catastrophic" and "instantaneous" were not words that geologists were accustomed to using when describing changes in climate.

What could have caused such an event? The Greenland team ruminated on the possibilities. Could ice surges from the Antarctic have triggered such swift changes? It seemed unlikely that events in an ocean at the opposite pole could transmit such changes so abruptly to northern Greenland. Could intense volcanic activity have caused it? No telltale volcanic layer showed up in the Camp Century ice at that depth.

Dansgaard noticed that the event was one of three similarly rapid climate collapses that were recorded deep in the Camp Century ice. All three had occurred during periods when orbital features diminished the radiation from the Sun. Yet only the last event, 73,000 years ago, ushered in an ice age, he observed. The mechanism may be a mystery, but the link to Earth's orbit was irresistible. Among the three abrupt changes, only the last was followed by 60,000 years of slightly lower summer midlatitude solar radiation.

As if setting an agenda, Dansgaard then posed a series of questions that would occupy many climate scientists for the rest of the twentieth century. "Were the sudden decreases in [oxygen isotope ratios] triggered by low insolation in the Northern Hemisphere, where extensive areas of land can be covered by ice? If so, the conditions for a catastrophic event are present today. . . . Or, are we faced with more or less accidental events, such as ice surges or intense volcanic activity, that trigger a full glaciation, if the insolation conditions favor such development? Is man's present activity equivalent to such [an] accidental event?"

Most of the conference participants followed the Milankovitch curves and the contemporary cooling trend and suggested that the end of the current warm period and the onset of a new ice age was a strong possibility. Whether it would come in a few centuries or a few thousand years was anybody's guess. Uncertainties about the impact of what Dansgaard called "man's present activity" hung over the proceedings. Cesare Emiliani observed that scientists could not predict the amplitude of change it might provoke or even the direction in might take. But his own research into the ice age problem had established a new and important fact, "that temperatures as high as those of today occurred for only about 10% of the time during the past half million years." After 10,000 years of warmth, the system was in a state of "precarious environmental balance" that makes man's interference "extremely critical."

Everyone agreed that they needed to know more. The profile of change in the Camp Century ice and the increasingly sharp focus emerging from the laminations in ocean sediment cores were revealing a truly baffling history of climate. Although new instruments and ingenuity were shedding new light on the subject, the fundamental problem of the past, the ice age problem, had not been resolved very satisfactorily. Like a microscope turned on a drop of pond water, the new records revealed details of events that were totally unexpected. Rather than answering old questions, the increasing refinements of vision raised entirely new mysteries.

Along the way, a new science of paleoclimatology had been born. Some of it was fashioned from the old disciplines of geology and astronomy, to be sure, but other elements were spanking new, invented in the richly financed postwar laboratories of nuclear technology. Engineering breakthroughs made possible the exploration of climate records in ice sheets and ocean sediments that opened up new archives of climate history. And human ingenuity was devising new ways to tease information from proxies in fossil relics of epochs past.

Many theories had been proposed. Leads had been followed down a lot of "blind alleys." Breakthroughs had been celebrated prematurely. Earth does not reveal its ways willy-nilly, especially the ways of its climate. Besides, that's the way it goes in science. As historian of science Spencer R. Weart has observed, "Every great scientific paper is written at the outside edge of what can be known, and deserves to be remembered if there is a nugget of value amid the inevitable confusion."

6

INSTABILITIES

"It got cold, and they died."

At the beginning of the fourteenth century, the northernmost outpost of European civilization was a long-lived and well-established settlement. In a medieval world torn by political and cultural turmoil, by war and want, the Greenland colony of the Old Norse had for centuries stood as a bastion of conservative stability. Its Scandinavian farmers were preserving a way of life that had endured since the colony's founding in 985 by Erik the Red. It was small in numbers, with perhaps 6,000 inhabitants in all in two settlements, but still it looked like a permanent feature of European geography.

The Norwegian Viking's renegade ways were quickly displaced by an uncompromising social structure built on European ideas of respectability and wealth. While Erik held to the ancient Norse pantheon, his wife, his son, and his fellow colonists quickly converted to the new Christian faith. They sent a live polar bear to the King of Norway to induce him to send a bishop. They endowed the bishop with a large, prosperous farm and built a cathedral, churches, and a nunnery, although by the beginning of the fourteenth century, the bishops had found it more convenient to govern the colony from Rome.

The Norse farmers had developed an economy around the attributes of a landscape still dominated by glaciers and a climate controlled by the North Atlantic. First and foremost, they were livestock and dairy farmers, practicing a traditional husbandry that was common to Scandinavia. Through the cold winter, they stabled their animals indoors. In spring and summer, every patch of viable ground was used for pasture to sustain livestock through the growing season and to harvest for winter fodder. In the warmer seasons, they fished for cod and traveled up the western coast to hunt seals and caribou for food. They hunted walrus for fur and for their ivory tusks, their most important export to Iceland, Ireland, and Norway.

By the beginning of the fourteenth century, the workings of Earth already were turning against the Norse farmers of Greenland. To the north, pack ice was growing, threatening to close off the sea route to the summer hunting grounds. Increasingly frequent storms and encroaching sea ice were making the route between Europe, Iceland, and Greenland more and more dangerous, threatening to isolate the colony. Even for such a hardy people accustomed to severe winters, life was becoming intolerably hard. The old Norse farmers exhausted their supplies of food and fodder before the end of winter. Depleted and demoralized, they watched their young and elderly die first. In desperation, in violation of ancient Norse law, they started eating their dairy cattle. More desperate still, finally they ate their hunting dogs. Starting with the northernmost Western Settlement about 1350, they abandoned their farms. A century and a half later in the Eastern Settlement, nearly 500 years of European settlement came to an end and the Norse people disappeared from Greenland.

These farmers were not the only medieval people to feel the lethal power of changing climate. And advancing ice was not the worst of it. The change that began in the fourteenth century altered history in Europe and elsewhere for the better part of five centuries. Around the world, one weather-related disaster followed another. In Asia, torrential rains in 1332 led to floods that devastated China, causing the deaths of several million people. From midcentury, China lost 40 percent of its population, 40 million people, within 100 years. Along with the change in weather came another calamity. A virulent new strain of bubonic plague—the Black Death—emerged, and in the following decades, waves of the disease swept through Asia and Europe. In England and Europe, the winter and the long summer and autumn of 1316 saw almost constant rains that led to widespread crop disasters, spreading famine and disease. Starving peasants resorted to cannibalism.

The inviting climate that first encouraged the Europeans to settle Greenland was described by early paleoclimatologists as the Medieval Warm Period; the change that finally drove the Old Norse out is called the Little Ice Age. Although it is generally true that the several centuries before the fourteenth century were warmer than the several centuries after it, these names imply a climate record that is neater than the real pattern of the past. Conventional standards of measure reveal variations during the past 10,000 years as more subtle than the sharp lurches in temperature and precipitation that took place during the ice ages of the past. In fact, climate scientists used to think that the changes during the 10,000 years of the modern era, the Holocene, were too minor to be interesting subjects of research. But the methods of modern paleoclimatology that exposed the reality of abrupt climate change also illuminated the powerful role of climate variation in the record of human events.

A new level of detail was detected by Chet Langway even in the first Greenland ice climate profile, from the 411-meter Site 2 core, which he developed in the 1950s. Looking back 10 centuries to the year 934, Langway saw trends in snow accumulation and temperature that suggested snowfall increases as temperature rises. If this is so, Langway reasoned, "then the record in the deep core may be read as meaning that the temperature as well as the accumulation in AD 934 was similar to that of today. From AD 934, a slow decrease in temperature and accumulation appears to have begun which reached a minimum around the late 18th century; thereafter a gradual warming and increase in precipitation occurred [and are still occurring] today."

As the climate record was understood in more detail, the settlement histories of both Iceland and Greenland would be interpreted in new ways. Langway and Willi Dansgaard came across more detail with their oxygen isotope analyses of the 1,400-meter Camp Century core. In 1970, they reported on their study of ice that fell as snow during the last 800 years, although this study was not focused on historical events. They were looking for patterns, for "systematic oscillations" that might give clues to the causes of change and help predict future variations. Only parenthetically they observed "the 'little ice age' 1600-1730; and the long-lasting minimum in the 15th century, which might have been a contributory reason for the dying out of the Norsemen settlements in Greenland."

In 1975, Dansgaard and his Geophysical Isotope Laboratory colleagues attempted a more deliberate examination of the climate changes surrounding the rise and fall of Norse Greenland—a subject, he noted, that "has fascinated generations of historians, maybe because it is the only example of a well developed European society being completely extinguished in historical times." In an article in the journal *Nature*, Dansgaard attempted to correlate oxygen isotope variations in this ice core profile "with some well documented important events in the early Icelandic and Norse societies." This study represented the first attempt to match paleoclimate proxy information to recorded human history. Dansgaard and colleagues found in the polar ice new information about both the beginning and the end of the Norse colony.

Searching for an ideal site for its next deep-core operation, the Greenland Ice Sheet Project team, using its thermal drill, had been sinking test cores in a number of sites around the ice sheet. The summer of 1974 had produced an especially promising core at a location called Crête, 10,315 feet above sea level at the crest of the ice sheet in central Greenland. In the spring of 1975, in the journal *Nature*, the GISP team reported the results of an oxygen isotope analysis of a 404-meter Crête core.

First, Dansgaard threw new light on the old explanations for the apparently contradictory names of the two North Atlantic islands. Iceland is definitely greener than Greenland. Greenland's landscape is dominated to a much greater extent by ice. Indeed, the anonymous author of a thirteenth-century Norwegian text known as the "King's Mirror" blamed Greenland for Iceland's ice. "As to the ice that is found in Iceland," he wrote, "I am inclined to believe that it is a penalty which the land suffers for lying so close to Greenland; for it is to be expected that severe cold would come thence, since Greenland is ice-clad beyond all other lands."

Dansgaard observed how Iceland got its name from a Norwegian farmer's failed attempt to settle the island in about 865, during a "short cold period" detected by the GISP team. The disappointed farmer, having lost his cattle in a severe winter, returned to Norway and told how he had watched a fjord fill with sea ice. Students of geography have long been amused by the account of the naming of Greenland in the medieval sagas. Centuries after the fact, a chronicler in Iceland maintained that the Viking Erik the Red chose the name as a subterfuge in order to entice more would-be settlers from Iceland. But the oxygen isotope evidence in the polar ice offered another explanation. The discovery of Greenland in 982 came "at the end of a warm period longer than any that has occurred since," Dansgaard reported. The name Greenland at that time "probably described a reality. . . . So, the drastic climatic change late in the ninth century may be part of the reason why Iceland and Greenland did not get the opposite names, which would have been more natural had they been discovered simultaneously."

No doubt, the cold fourteenth century marked "the beginning of the tragic end," he wrote, although the reasons for the abandonment of the Western Settlement around 1350 and the disappearance of the Norse people from the Eastern Settlement a century later were not clear. The oxygen isotope curve "suggests that the climate became so cold in the fourteenth century that years of famine must have occurred frequently, as was the case in Iceland, particularly in the 1370s. On the other hand, the cold made whaling a good business for European mariners who may have assaulted the Østerbygd [Eastern Settlement]—at least Eskimo tales blame pirates for the extinction of the weakened Norse society. Nor can it be excluded that the plague, which came to Iceland in 1402, was transferred to Greenland by pirates." While climate alone may not account for what happened in Norse Greenland, Dansgaard wrote, "the impact of climate on human life stands out more and more clearly, particularly where the margin of survival is slim, whether in the polar regions or in the tropics."

From polar ice taken 20 years later from a core near Crête, another investigator would be able to discern conditions during individual seasons in developing a picture of the climate stresses faced by the Norse in Greenland in the fourteenth century. In the 1990s, Lisa K. Barlow used the stable isotopes of hydrogen, the other atom in the water molecule, to construct the same sort of temperature signal that the GISP team and others developed with the isotopes of oxygen. Like the stable oxygen isotopes, the differences in the ratios of the heavier and lighter hydrogen atoms in the water molecules reveal differences in temperature. The lighter hydrogen isotope evaporates at a slightly lower temperature. A rising proportion of the heavier hydrogen isotope deuterium in a sample of ice core signals a rising temperature.

In 1997, Barlow reported that the climate record, built from the hydrogen isotope signal from an ice core in central Greenland, suggested that the Norse settlers experienced above-average temperatures for the first 6 to 12 years after landing in 985 and that this was the longest period of "mild" weather for centuries to come. And according to the new isotope record, the fourteenth century was the coldest period in the region in 700 years. In particular, Barlow identified a 20-year period of especially cool summers, from 1343 to 1362—the very time when the Western Settlement is thought to have collapsed. "Norse farmers relied as much on summer fodder production to last through the winter as on a timely end to winter conditions," Barlow wrote in 1997. "Thus a period of cold summers may have reduced grass production, making the subsistence of the Norse Greenlanders even more precarious."

To be sure, "it got cold, and they died," as an anonymous medieval author described what happened in Norse Greenland, but Barlow and others maintain that climate variation tells only part of the story. She is a member of a team of researchers from a variety of scientific and scholarly specialties who investigated the history of Norse Greenland. The subject has been a battleground for competing interpretations by generations of researchers. The new team is trying to settle these old academic disputes and gain a clearer picture of what really happened in the fourteenth century in the Western Settlement and 150 years later in the Eastern Settlement.

To anthropologist Thomas H. McGovern, the rigid character of Norse society left it vulnerable to change. As it happened, he notes, while the Norse were losing their battle with the elements, the native Inuit were winning theirs, flourishing in a more mobile and adaptive culture of Arctic hunting. Their toggled harpoons, which open on a hinge after piercing the flesh, allowed the Inuit to successfully hunt seals through the ice during winter. Unlike the wooden Norse boats, the seal skin Inuit kayaks could be lifted and easily moved over the ice jamming the fjords. Even their clothing of fur and skin was warmer and more practical than the Norse woolens that were woven according to European fashion. The Norse increased the proportion of seafood in their diet and rearranged the rooms in their houses to take advantage of the body heat from their livestock, but these changes were not enough. The judgment of history would be that in the face of changing climate these European farmers were stuck in their ways, in their fixed abodes, in their property rights, in their royal taxes and church tithes, and in their belief that the way of life that had sustained them for so long would pull them through the current bad weather. And their leaders were preoccupied with matters of status and wealth.

Our fascination with the extinction of the Norse people in Greenland may be more than academic. Something about it haunts our history in ways that other calamities do not. Maybe it is because we modern observers of the Greenland colonists' "failure to adapt" have more in common with the Norse than with the Inuit. They are figures in shadow that we faintly recognize. Are we not as heavily invested in the idea of climate stability? Are we not as fixed in our ways, just as devoutly faithful to our own cultural imperatives? Those seeking answers to questions such as these finally arrive at a philosophical divide. In 1984, the English climate scholar Hubert H. Lamb called attention to "a general resistance in mankind since the modern scientific and technological revolution to the idea that our way of life may be drastically affected by natural changes and fluctuations in the physical environment. This attitude may hold dangers for the future. Our forefathers had a very different view, based on bitter experience of disasters from famine and disease."

Resistance to the idea was easier in the days when our vision of the past was so rudimentary that the 10,000 years of the modern Holocene period looked climatically stable and uninteresting. Early in the twentieth century, it seemed to geographer Ellsworth Huntington that he was finding the heavy hand of climate everywhere he looked in the historical record. He so overstated the case for "climate determinism" that the subject quickly fell out of fashion among scholars. Now polar ice profiles and other powerful climate proxies such as tree ring analyses are forcing interesting reevaluations of the past 10,000 years. The conclusion is hard to ignore: It has been a bumpy ride.

At the end of the fifteenth century, at a time when the home country and the Mother Church seem to have been entirely in the dark about the fate of the Greenlanders, another European explorer was setting sail on a new voyage of discovery. Christopher Columbus did not need to be warned about the hazards of the northerly route across the North Atlantic; but what he did not know was that while his more southerly route would avoid the treacherous northern ice, it would expose him to another powerful threat. The phenomenon of a hurricane was entirely unknown to the Old World. If European sailors of the day had seen the likes of the summertime Atlantic's marauding tropical storms and hurricanes, none had lived to tell about them.

The earliest reports of the devastating effects of a hurricane came to us from Columbus. Three of the ships that accompanied his second voyage were wrecked by hurricanes on the north coast of Hispaniola in 1494 and 1495. What if he had encountered such a storm on his first voyage and been caught in the open sea in the three tiny caravels in 1492? A conspicuous chapter of history might have gone quite differently. Of the Hispaniola experience, he wrote that "nothing but the service of God and the extension of the monarchy" could induce him to expose himself to such dangers.

Old World sailors had no idea where these monster storms came from or, more importantly, when they were likely to appear. The Spanish fleet would assemble in Havana each spring in preparation for the voyage to Spain. Customary fiestas and other activities there would extend their time of departure into the height of the hurricane season. In the century before the push for colonization, hurricanes haunted the gold-hungry Spanish, regularly sinking treasure-laden galleons plying Columbus's route. Just as sailing vessels do today, they set a course for the New World from the Canary Islands to Guadalupe, and they returned in a route north of the Sargasso Sea toward the Azores. Ponce de Leon encountered two shipwrecking hurricanes within a span of 13 days in the summer of 1508 off the coasts of Hispaniola and Puerto Rico. A hurricane in October 1525 sank the first vessel sent to Mexico by Hernando Cortes, who had discovered gold there. In 1559, a hurricane wrecked the Spanish expedition sent to establish a colony near what is now Pensacola, Florida.

The climates of North America and their weather extremes were abiding mysteries to Europeans. Seeing only the likeness of their latitudes between the Old World and the New, the savants erroneously supposed a likeness of climate. Because they were in the same latitude, the French expected the lands of New France in Newfoundland to enjoy the same climate as central France. And New England, 10 to 12° South of old England, was expected to have milder weather. Instead, as the American weather historian David M. Ludlum observed in 1967, "winters were more severe than anticipated, and summers much hotter than anything experienced in the homeland." Later, of course, New England would be seen as a region controlled by continental climate factors rather than the maritime influences of the British Isles.

Settlers were victims of more than just ignorance. New World colonies became pawns in the political struggles of Europe. To the English, Roanoke Island looked like a strategic outpost for raids on Spanish shipping out of Florida. Sir Walter Raleigh estimated that these rewards of privateering would finance the whole colonizing enterprise. Royal treasuries and private riches were at stake. The lavish descriptions of the wealth and wonders of the New World came from witnesses who pandered to the expectations of royal courts and private sponsors. The natives were friendly, food was plentiful, and land was free for the taking. The supposed salubriousness of the climate was just part of the pitch.

The first French settlers in North America felt "greatly deceived" by the descriptions they had been given of the climate of the new land. One writer described the "temperature of the climate" of Maine as "so temperate, as it seemeth to hold the golden mean." On arriving in Virginia, John Smith wrote that "heaven and earth never agreed better to frame a place for man's habitations." That such a description would come from the first leader of Jamestown in 1607 may tell more about the scruples of Sir Walter Raleigh and the Virginia Company than about conditions in Virginia at the time.

Despite Raleigh's attempts to conceal it, by 1607 most everyone in England had heard the truth about what heaven and earth had wrought upon his first efforts to establish a colony in Virginia. At Roanoke, just 130 miles south of Jamestown, three attempts had ended in one failure after another. Most recently, more than 117 English subjects—including soldiers and planters, 17 women, and 11 children—had vanished from Roanoke after 1587. The fresh mystery of their fate must have haunted some dreams in Jamestown.

The "Lost Colony of Roanoke" has been the subject of speculation by generations of historians, archaeologists, and other researchers. Explanations have been hard to prove, of course, but easy enough to come by.

At a time when the colony would critically need food from nearby natives, relations with neighboring tribes had already been poisoned. The first English colonizers, among them bellicose privateers and military men, were openly at war with the natives when they quit the place and returned to England with Sir Francis Drake in the summer of 1586. When the second group of colonists arrived in July 1587, they found that all 15 soldiers who had been left behind to secure the fort had been killed. Retaliating for this offense, the new colonists burned a nearby village and mistakenly slaughtered a group of friendly Croatan natives.

It could also be said that the colony was sacrificed to the greater good of England. In August, when the colonial governor, James White, returned to England for new supplies, he found that all ships were commandeered for the battle against the Spanish Armada in 1588. By the time he finally was able to return in 1590, the colony had been abandoned and White could find no trace of the settlers, including his daughter and newborn granddaughter. This is the traditional explanation for the disappearance of the Roanoke Colony.

In 1998, however, a team led by climatologist David W. Stahle and archaeologist Dennis B. Blanton developed a new and terribly plausible explanation. Like the Old Norse of Greenland, these earliest English settlers were unlucky victims of rapid climate changes. Stahle took tree ring cores from 800-year-old bald cypresses taken from the Roanoke Island area of North Carolina and the Jamestown area of Virginia, reconstructed precipitation and temperature chronologies, and concluded that the timing of the early English colonizing efforts could not have been worse.

The settlers of the Lost Colony landed at Roanoke in the summer of the most extreme growing-season drought in 800 years. "This drought persisted for 3 years, from 1587 to 1589, and is the driest 3-year episode in the entire 800-year reconstruction," Stahle and his colleagues reported in the journal *Science*. A map shows that "the Lost Colony drought affected the entire southeastern United States but was particularly severe in the Tidewater region near Roanoke." The authors surmised that the unsuspecting Croatan who were shot and killed by the colonists that summer were scavenging the abandoned village because they already were short of food.

"The tree-ring reconstruction also indicates that the settlers of

Jamestown Colony had the monumental bad luck to arrive in April 1607, during the driest 7-year period in 770 years," wrote Stahle. The ragged story of the early settlers of Jamestown is better documented of course, although in seventeenth-century England, Raleigh and his commercial cohorts were loathe to make it known. Like their Roanoke colleagues, the Jamestown settlers did not know what the Algonquin natives knew about the drought and its effect on their food supplies and on the quality of water in the York and James Rivers in the Tidewater peninsula of Chesapeake Bay. Most of them would not survive their ignorance of these things—and their bad luck.

First came the long hot summer of 1607. Having arrived in May, weakened by their long sea voyage, many fell victim to fevers from diseases that prevail in heat. The warm autumn was followed by terrible cold. When the first supply ship arrived in midwinter, it found only 32 of the original 105 settlers still alive. In January the entire town was destroyed by fire. Captain John Smith later wrote: "Many of our old men diseased; and [many] of our new, for want of lodging, perished."

Soon relief was on the way. Food and other supplies would come from the Virginia Company in England. How many sick and starving emigrants must have prayed for its arrival. The Virginia Company in London provisioned nine vessels that left England on June 2, 1609, under the command of Admiral Sir George Somers in the flagship *Sea Adventure.*

Sailing south of the tropic of Cancer, the seamen found themselves sweltering in the humid heat. Aboard the vessel *Blessing,* Captain Gabriel Archer reported that "by the fervent heat many of our men fell sick" and from just two ships, 32 bodies were thrown overboard.

On July 24, as the relief flotilla labored within a few days of the shores of North America, the ships sailed into the maw of a hurricane. Aboard the *Sea Adventure,* William Strachey recalled it as "a dreadful storm," violent beyond imagining. "Our clamours drowned in the winds and the winds in thunder," he wrote to a lady friend. "The sea swelled above the clouds and gave battle to the heavens." After four days of such violence, its 140 men exhausted by constant bailing, the *Sea Adventure* found itself sailing alone. Admiral Somers spotted land, the so-called Isles of Devils, known to mariners since 1511 but avoided for their rings of treacherous reefs and reputation for stormy winds. Miraculously, all aboard the *Sea Adventure* survived the landing, although the vessel was damaged beyond repair.

Meanwhile, seven battered ships in the Virginia Company flotilla managed to stay their course and make Jamestown by late August. Rather than relief, however, the 400 surviving passengers, sickly and weakened from the terrible voyage, instead brought greater burden. Much of their food supply had been thrown overboard when the storm threatened to sink the vessels. The rest was waterlogged and ruined. Another crisis winter, including another especially cold month, followed. What happened then John Smith retold in 1624.

"Nay, so great was our famine, that a savage we slew, and buried, the poorer sort took him up againe and eat him, and so did diverse one another boyled and stewed with roots and herbs," Smith wrote. "And one amongst the rest did kill his wife, powdered (salted) her, and had eaten part of her before it was knowne, for which hee was executed, as hee well deserved. . . . This was that time, which still to this day we call the starving time, it were too vile to say, and scarce to be beleeved, what we endured."

The men and women of the *Sea Adventure*, meanwhile, found the uninhabited Isles of Devils "abundantly fruitful." They feasted on wild pigs that had been left by the Spanish and other plentiful game and enjoyed a temperate climate. Strachey called it "the richest, healthfullest and pleasing land . . . as man has ever set foot upon." From the accident of shipwreck, a permanent settlement was established. The islands were known for a time as Somers Islands, although they were later named Bermuda in honor of the Spanish seafarer Juan Bermudez, who first plotted them on a map. Still Bermuda's coat of arms features the wreck of the *Sea Adventure*, and still the island celebrates Somers Day on July 28.

The healthy and industrious settlers fashioned two smaller

boats from the wreckage of the *Sea Adventure*, and the following summer most of them finally made their way to Jamestown, where they found 60 gaunt survivors of the starving time.

"Viewing the fort," wrote William Strachey, "we found the palisades torn down, the ports open, the gates from off the hinges, and the empty houses (which owners had taken from them) rent up and burnt, rather than the dwellers would step into the woods a stone's cast off from them to fetch other firewood. And it is true, the Indians killed as fast without, if our men stirred but beyond the bounds of their blockhouse."

The saga of the Bermuda shipwreck of the *Sea Adventure* became well known in London. Passenger Sylvester Jourdain wrote a lengthy account that was a literary hit in 1612. Also widely circulated was a detailed letter by Strachey, who happened to be an investor in London's Globe Theatre, where William Shakespeare was a partner. Soon thereafter, Shakespeare wrote one of his finest plays, a story about disaster at sea, a shipwreck, and a storm. He called it *The Tempest.*

7

CONVERGENCE

Abrupt climate change finds a theorist

The new methods of reconstructing climates past not only have inspired reinterpretations of history but have also cast the future in a new light. They have transformed the study of ancient climates into the study of the very nature of climate—the ways it distributes heat and water around the planet and the dimensions and pace of its changes.

Nothing stimulated the field of paleoclimatology more than the reports in the early 1980s by Swiss and French physicists and climatologists of their analyses of the composition of the bubbles found in polar ice cores. Researchers suddenly realized something that many had not even dreamed possible: Not only was polar ice an archive of the isotopic footprints of past climates, it was a storehouse of the climate system's most ephemeral feature—the air itself. The bubbles in the ice were tiny time capsules of "fossil" atmospheres that were trapped inside the glacier when the coarse old snow turned to solid ice about 100 meters down. They could be assigned ages, and unlike proxy evidence, the chemical composition of this trapped air could be analyzed directly.

These technically demanding geochemical analyses were accomplished by the Swiss physicist Hans Oeschger and the French physicist Claude Lorius. Independently of one another, Oeschger, analyzing Greenland ice, and Lorius, analyzing Antarctic ice, measured the level of carbon dioxide in the air bubbles.

For the first time, scientists from a broad range of specialties began to recognize polar ice as a source of climate information that was timely and global in scope. Carbon dioxide, a gas that is well known for its radiative "greenhouse" properties even in minuscule concentrations, remains in the atmosphere hundreds of years. Long-lived gases are dispersed evenly around the world by the mixing actions of the atmosphere. Even scientists who were skeptical of the oxygen isotope analyses of the ice and their evidence for abrupt change realized that the air pockets held clues to what was becoming the most compelling environmental question of the day: How will Earth's climate respond to industrial pollution?

The answer to this critical question hinged on the chemical composition of the atmosphere and on its history. In the face of rising global temperatures and mounting evidence, predictions of climate scientists had radically changed. Concern about Earth's natural long-term orbital variations forcing it toward a new ice age had given way to a vision of global warming provoked by human pollution from the burning of hydrocarbon fuels. Driving this change in thinking were three sets of critical data that came together within a decade.

In 1973, a landmark study by Charles D. Keeling of the Scripps Institution of Oceanography reported the first direct measurements of changing carbon dioxide concentrations in the atmosphere. Measured from his laboratory atop Mauna Loa, Hawaii, atmospheric CO_2 had climbed steadily from 312 parts per million in 1958 to 330 parts per million in 1972. These meticulous readings and the inexorable rise of the famous "Keeling curve" would become central to the issue of global warming.

Beginning in 1980, European polar ice laboratories produced the first convincing picture of how the composition of the atmosphere had changed through history and clues to the relationship of these changes to global temperatures. In 1980, researchers reported results of the first analyses of an Antarctic ice core that measured the concentration of carbon dioxide gas in the atmosphere during the nineteenth century. In these "preindustrial" times, scientists reported, the atmospheric CO_2 concentration was between 280 and 290 parts per million.

In 1982, the other shoe dropped. European labs reported that about 20,000 years ago, during the coldest depths of the last ice age, the atmosphere contained about 30 percent less carbon dioxide, about 200 parts per million.

"The discovery of natural oscillations in greenhouse gases from fossil air trapped in polar ice ranks as one of the most important advances in the field of climate and earth science," wrote geologist Thomas M. Cronin, author of *Principles of Paleoclimatology*, a leading text on the subject. Together with Keeling's measurements, he wrote, "These discoveries about natural and human-induced fluctuations in potentially climate-altering atmospheric gases sent shock waves throughout the paleoclimate community that still reverberate."

Now the relevance of the past to the present was no longer a subject of conjecture. At universities and government laboratories, new technologies were applied to the mysteries of the ice. Old institutional barriers gave way as specialists who had been only vaguely aware of one another's work found themselves in long conversations about the surprising results from Bern and Grenoble. Among a whole generation of young earth scientists, polar ice was becoming famous. For the first time, paleoclimatologists occupied the same conference rooms as scientists who were focusing on the increasing concentration of greenhouse gases in the contemporary atmosphere. And along the way, almost incidentally, a larger and more receptive audience heard about the Greenland ice cores and their evidence for abrupt climate change.

Comparing the Dye 3 core with the Camp Century ice and with new results from lake sediments in Switzerland, Willi Dansgaard and Hans Oeschger had identified 24 episodes of abrupt change during the last ice age. These times of sudden warming took on a characteristic, somewhat straight-sided rectangular shape in the profile of the ice cores. Whatever their cause, the switches in oxygen isotope readings apparently were recording the same sequence of events. Temperatures shot up to a level that indicated a distinctly milder climate, although not as mild as the modern climate of the last 10,000 years, the Holocene. Fairly soon after this spike, temperatures would begin an irregular slide back from this new level of warmth. Then, after several centuries, more or less, they would plunge back just as suddenly to ice age cold.

In the minds of these ice core researchers, a new conception was taking shape. The sharp, erratic changes seemed to show a climate that was shifting in and out of two distinct modes or states of operation, like a badly driven jalopy lurching between gears. In a 1984 paper, Oeschger, Dansgaard, Chet Langway, and others observed that these events "can be linked to oceanic changes, namely advances and retreats of cold polar water in the North Atlantic Ocean." The southward advance of polar water coincides with colder climate in Europe, they noted, and its retreat northward with a warmer Europe. "This strongly suggests that the continental climate shifts were an effect of changes of the surface conditions of the North Atlantic Ocean." But researchers were more inclined to think of the oceans as reservoirs of "thermal inertia" and climate stability rather than agents of sudden change. What could cause the ocean surface conditions to shift so rapidly?

No one in climate science was more interested in this question or better prepared to consider it than the American geochemist Wallace S. Broecker. Since the 1950s, Broecker had been following two research paths—paleoclimatology and ocean circulation. About all these subjects seemed to have in common was the fact that both were amenable to investigation by the new technology of radioactive isotopes of carbon. As a graduate student in charge of Columbia University's new radiocarbon laboratory, Wally Broecker was among the first researchers to use ^{14}C dating techniques to explore both subjects.

In 1957, his doctoral thesis described the use of ^{14}C dating techniques in both fields of research. Analyzing samples obtained with a 200-liter ocean water sampler developed by oceanographer Maurice Ewing, director of the Lamont Geological Observatory, Broecker was able to estimate the ages of volumes of seawater taken from different depths. From this data emerged a picture of the pattern and pace of ocean circulation. Inspired by radiocarbon dating of samples he had taken in Nevada from caves around a prehistoric lakebed, he also included a chapter entitled "Evidence for an Abrupt Change in Climate 11,000 Years Ago." As a professor at Columbia and a researcher at Lamont, Broecker had spent the next 20 years developing ways to measure the rate of circulation of the oceans and techniques to correlate various far-flung clues to climate events that marked the transition from the ice age some 11,000 years ago.

When Oeschger and Lorius reported that the ice age atmosphere contained a third less carbon dioxide, Broecker pounced on the results. While there had been "much discussion about the influence of the anthropogenic 'greenhouse' gases on future climate," Broecker wrote, until the results from the polar ice bubbles, researchers had "little of substance" to link paleoclimate to paleochemistry. "I realized right away that the driver for these changes must reside in the oceans," he recalled. After 20 years, the link between atmospheric CO_2 and ice age temperatures still has yet to be satisfactorily explained; "the driver" has yet to be identified. Ironically, even though, as Broecker put it, "the prize has yet to be grasped," his fascination with the subject led to his most famous scientific contribution.

All of his years of research into the chemistry of the ocean and the mysteries of paleoclimatology came together in 1984 as he sat in a lecture hall at the University of Bern, watching and listening to Hans Oeschger puzzle over the pattern he was seeing in the Greenland ice core. Years later, Broecker would recall the defining moment.

Oeschger was showing a graph of the Dye 3 climate profile from Greenland that depicted the numerous abrupt warmings during the last ice age that would become known as "Dansgaard-Oeschger events." The graph, which showed atmospheric CO_2

levels changing along with temperatures, suggested to Oeschger a system jumping back and forth between two different modes of operation. Broecker recalled: "Realizing that the CO_2 jumps required changes in the operation of the ocean, I said to myself, 'Could the key lie in a turning on and off of deep water formation in the northern Atlantic?' And as quick as that, my studies in oceanography and paleoclimatology merged." Again, Broecker's theoretical instincts would prove to be better than the underlying data. Researchers lost confidence in the CO_2 readings that seemed to rise and fall along with the Greenland warming events during the last ice age, especially after they failed to appear in the Antarctic ice. By then, however, Broecker was on his way to developing the first enduring explanation of abrupt climate change.

In 1985, in a defining article in the British journal *Nature*, Broecker took the subject from the sidelines and onto center stage in paleoclimatology. In the hands of its leading theorist, the focus of the science and its research agenda were about to be given a new look. Different events were going to be seen on a different timescale. The ocean sediments and their long but blurry geological reach, so helpful to the investigation of ice age rhythms, were going to have to make room for the polar ice cores and their shorter, more finely detailed mysteries of rapid change. With coauthors Dorothy M. Peteet and David Rind, Broecker rephrased the question posed by Oeschger, "Does the ocean-atmosphere system have more than one stable mode of operation?"

Broecker pointed out that the Greenland ice cores detected changes during the last ice age that hadn't been seen before—"many brief events during which climatic conditions returned about halfway to their inter-glacial state." While such fine detail would be lost in the typical sediment core taken from the open ocean, he noted, this did not mean that the ice cores and the sediment cores were really in conflict with one another. "However, as these idiosyncrasies of the ice-core record were not seen in other records, the initial temptation was to pass them off as climate 'noise' without global significance. A rapid succession of findings has since changed this view of the noise, now the focus of much interest."

The Dye 3 results, reported in 1981, had confirmed that the wiggles in the oxygen isotope profile that first showed up 10 years earlier in Camp Century ice were real warming events. The rapid changes also showed up in the ice core analyses of other chemical proxies—in dust and in concentrations of wind-blown aerosols of sulfate, nitrate, and chloride. And the oxygen isotope evidence for the most recent rapid event in both cores also appeared in calcium carbonate analyses from lake sediments in Switzerland, supporting Willi Dansgaard's view, expressed years earlier, that the Greenland ice and the European pollens were recording the same rapid cooling event—the Younger Dryas.

With this paper, Broecker began to describe the Younger Dryas as a climate event that was larger in scope and more consequential than his colleagues might assume from its obscure origin in the fossil pollens of Scandinavia. Signs of the dramatic climate tumble back toward ice age cold also had turned up in Spain, northern Italy, and northern Canada, although not in the United States. It may be a regional climate phenomenon, limited to the North Atlantic, but "the situation may be more complicated." Although the Younger Dryas wasn't detected in the Antarctic ice taken from a core at Byrd Station, there was evidence for such an event in the mountains of South America and in New Zealand.

This convergence of Broecker's research paths inspired a sustained effort to develop a theory of abrupt climate change—in particular, to explain the Younger Dryas. Broecker sketched a mechanism that described an intricate interplay between the movement of water vapor in the atmosphere and the transfer of heat in the oceans. The scenario linked atmospheric processes to a global interconnected system of large-scale ocean currents. Oceanographers call this system the *thermohaline* circulation. This joining of the Greek words for heat and salt signifies that the currents are driven by variations in temperatures and levels of salinity. In 1961, the famous oceanographer Henry Stommel had proposed that this global circulation has two stable states.

North Atlantic Ocean water is saltier than North Pacific Ocean water. The North Atlantic gives up more fresh water to evaporation

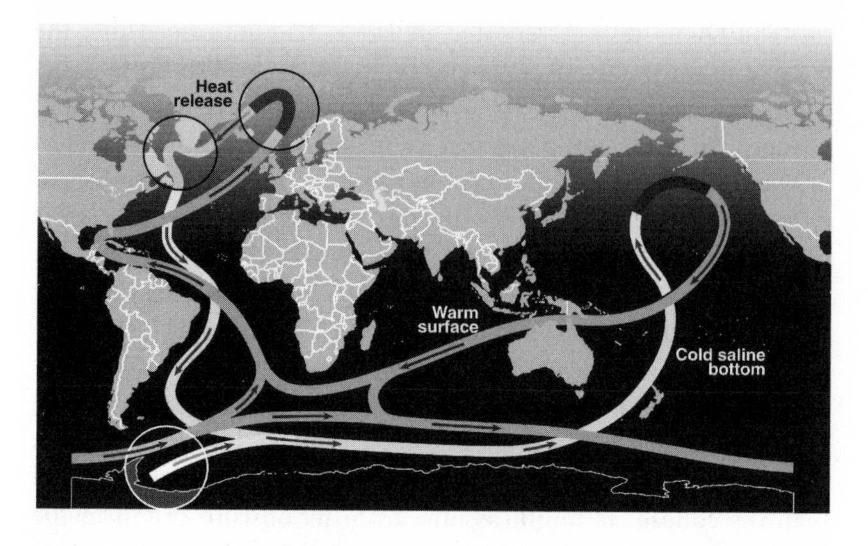

The Great Ocean Conveyor

First drafted by Wallace S. Broecker, this diagram illustrates how differences in water temperature and salinity transport water—and heat—among the world's oceans. Notice the key role of the North Atlantic as an important "cog" in the conveyor system. In two critical areas, where heat is released to the atmosphere, surface water becomes more dense, sinks to the ocean depths, and begins flowing southward. Reprinted from National Research Council, *Abrupt Climate Change: Inevitable Surprises*, National Academy Press (2002).

than it receives through precipitation and runoff from adjoining rivers. The opposite is true of the Pacific. At the same time, the North Atlantic's surface water is warmer than the Pacific's at the same latitude. As Broecker wrote, "Water is being distilled off the 'warm' Atlantic and condensed on the 'cold' Pacific" by the storms and winds of the atmosphere. Because of this constant freshening, the North Pacific's waters do not develop the same layering of different densities. Cold water from its great depths wells up along its continental margins and in a great clockwise gyre flows back over its surface toward the Equator. In the North Atlantic, the pattern of circulation is more distinctly vertical. In the North Atlantic, the big warm surface currents, the Gulf Stream and the North Atlantic Drift, carry heat from the Tropics toward the pole and the cold abysmal currents carry excess salt southward from the Atlantic to the Pacific and Indian oceans.

Broecker focused on a critical juncture in the system—the far North Atlantic, the Labrador and Nordic seas, where water from the surface encounters the Arctic westerlies off Canada. Now cold and salty, the current sinks to the abyss, forming the dense southward flow that oceanographers call North Atlantic Deep Water. The evaporative cooling of the surface water transfers from the ocean to the atmosphere an amount of heat that Broecker estimated to be equivalent to 30 percent of the Sun's warmth that far north. Northern Europe, which is meteorologically "downstream" of this process, enjoys a climate that is warmer than other regions at such high latitudes.

Using the new computer model of the general circulation of the atmosphere, developed at the Goddard Institute for Space Studies in New York, Broecker, Peteet, and Rind built the most persuasive case yet for the role of ocean circulation in rapid climate change. They noted that, according to a variety of evidence from deep-sea sediments, the formation of this North Atlantic Deep Water "was reduced greatly" during the coldest depths of the last ice age. But what of the abrupt warmings, Broecker wondered: "Is it possible then that the brief warm events recorded in the ice cores represent periods during which the glacially weakened northern Atlantic deep-water source was rejuvenated?" The scientists couldn't test that question directly on the computer model without more evidence of the geographical distribution of the climate impacts of warm events. But they could test the opposite circumstances—the last big rapid event, the sudden cold of the Younger Dryas.

The researchers plugged into the Goddard Institute model the colder sea surface temperatures estimated for the North Atlantic during the depths of the last ice age. The computer generated cooler temperatures across Europe and far northeastern North America in a pattern that looked like the pollen record of the Younger Dryas. Of course, it wasn't proof that a shift in the North Atlantic's circulation had provoked the cold of the Younger Dryas or the ice age warm events. But as scientists are fond of saying when using computer models, the results were consistent with the evidence.

To Broecker it was tempting to speculate that Oeschger's two modes represented two states of operation of ocean circulation. During the ice ages—the glacials—the North Atlantic was cold because the circulation was weak or even reversed, he supposed, and during the warm interglacials the North Atlantic was warm because the circulation would be strong. During the "Oeschger oscillations," as he called them in this paper, climate seemed to settle temporarily at some halfway point before slipping back into the ice age. Broecker tentatively outlined a scenario of abrupt warmings and cold snaps driven by changing ocean temperatures, ice cap melting, and shifting seawater salinity.

As Broecker noted, scientists had generally assumed that the climate system's response to "any gradual forcing will be smooth," but if Oeschger was right about the system having more than one stable mode of operation, "then the situation is more complex." If Oeschger was right, atmospheric computer models were going to have to become more sophisticated and incorporate the physics of the ocean to allow researchers to explore the more complex interactions in the climate system, and scientists were going to have to learn more about its components. If Oeschger was right, the great challenge facing climate science—anticipating future change—not only was more complicated, but its solution might well be more urgent. Broecker began to wonder about the implications of ever-increasing concentrations of carbon dioxide in the atmosphere. Are sudden mode switches likely in such a future?

The information was tenuous, he acknowledged, and thinking in terms of these abrupt mode changes was not going to be easy, but "we must begin to explore this alternate track." The new approach meant that scientists were going to have to design better computer models, achieve a better understanding of the climate system's various parts, and "extract all possible information from the paleoclimatic record." More polar ice cores were going to have to be drilled. More money was going to have to be spent. More scientists were going to have to be involved. "Unless we intensify research in these areas," Broecker wrote, "the major impacts of CO_2 will occur before we are prepared fully to deal with them."

In the mid-1980s, Broecker began a long, fruitful exploration of "this alternate track" of abrupt climate change. A transformation was under way. While other scientists devised new ways to tease information from archives of climate in glacial ice and ocean sediments and in tree rings and sea corals, Broecker would assimilate the data and give it theory—a physical mechanism that explained the data. The polar ice cores and the refinement of methods of analyzing ocean sediments were bringing into focus new features of climate with new causes and effects, and a new terminology was taking shape, much of it coined by Broecker.

To distinguish the abrupt episodes of relatively short-term change from the variations that marked the grand ice age cycles, abrupt changes became known as "millennial-scale" events because they seemed to last about a thousand years, more or less. The term itself is a bit of geological artifact, in a sense. To scientists accustomed to thinking in units of geological time, calling something "millennial scale" was meant to imply brevity.

Broecker coined the name "Dansgaard-Oeschger" oscillations, or D-O events, for the series of sudden warmings through the last ice age. The term recognized Willi Dansgaard's pioneering oxygen isotope analyses of the ice core that first revealed the temperature changes, as well as Hans Oeschger's landmark interpretations of these events.

Before long, the terminology of the new science would include other coinages for other abrupt climate oscillations. The young German researcher Harmut Heinrich, examining layers he detected in Atlantic Ocean sediments, identified a series of sudden plunges to especially cold temperatures during the last ice age. The sediment layers were composed of debris scraped off by the grinding of the Laurentide ice sheet in Canada and rafted far across the North Atlantic by armadas of icebergs. Broecker named these cold outbreaks of icebergs "Heinrich events."

But the centerpiece of the new science remained that mysterious stab of cold that so dramatically interrupted the 4,000 years of warming from the last ice age, the Younger Dryas. On land and at sea, evidence for this last and most accessible rapid climate change 11,000 years ago continued to accumulate, enlarging its reach and its global significance. By 1985, Broecker had satisfied himself that an abrupt shutdown of the ocean circulation in the North Atlantic was responsible for ice age cold suddenly spreading back across Europe. But what would cause such a catastrophe?

In the mid-1980s, Broecker put the pieces of the puzzle together. His explanation of the Younger Dryas changed contemporary thinking about the character of Earth's climate. The message from the Greenland ice cores was clear: the behavior of climate during the last ice age was nothing like the epoch of slumbering stability that was commonly accepted. Nor were the great ice sheets it formed subject only to ponderous advance and decay in concert with the variations of Earth's orbit. Just as the Camp Century and Dye 3 cores implied, sudden large swings between exceptional cold and exceptional warming marked 100,000 years of climate history. What Broecker brought to the science was the first plausible explanation for these events.

The Younger Dryas, Broecker said, was caused by the sudden massive flooding into the North Atlantic of meltwater from the Laurentide ice sheet that covered much of North America. Some 2,000 years of warming had led to the formation of an enormous lake across much of Canada, south and west of the world's largest glacier. About 11,000 years ago, the retreat of the ice closed off the lake's southern outlet through the Mississippi River basin into the Gulf of Mexico and opened a new channel eastward through the St. Lawrence into the North Atlantic. This sudden freshening of the surface water altered its density balance, preventing it from sinking to the ocean's depths and blocking the northerly flow of the warming current from the Tropics.

Climate scientists would spend years debating the details of this scenario and filling in the missing pieces. Broecker's bold line of thought drew widespread interest and brought a new focus to paleoclimatology. For the first time, climate scientists had a coherent explanation for abrupt changes, one that invoked a close interplay between processes in three realms of the climate system: the atmosphere, the ice sheets, and the oceans. And it brought together the evidence from three very different lines of investigation: ice cores drilled from the Greenland ice sheet, sediments plumbed from the seabed of the North Atlantic, and fossil pollen layers in the old bogs and lakebeds of northern Europe.

A new time dimension was being forced onto earth science, one remarkably close to the old catastrophist fantasies that geologists had fought so hard to disassociate themselves from earlier in the century. So much for the time-honored maxim of Aristotelian thinking: *Natura non facit saltum*—nature does not make leaps. When it comes to changing climate, it turns out that making leaps is *exactly* nature's way. In 1989, the time dimension was given a precise new scale by another study of the Camp Century and Dye 3 ice by Dansgaard, the American James W. C. White, and Sigfus Johnsen of Iceland. Examining the core sections representing the cold-to-warming transition that marked the end of the Younger Dryas, these researchers concluded that most of the "abrupt and radical changes" had happened in the span of only 20 years.

Like Hans Oeschger in Switzerland, Broecker was quick to relate the lessons of abrupt change to the growing dialogue about the climate impact of the rising atmospheric concentrations of carbon dioxide and other greenhouse gases. He sought to extend these new lines of thought to the broader community of climate scientists and to alert policy makers and the public to the disquieting new discoveries in the paleoclimate record. Broecker is an entertaining speaker and an acknowledged leader in his field, and everyone listened politely. But none of these audiences found his message particularly welcome, primarily because his news was not good.

Many fellow earth scientists, steeped in the conservative tradition of geological time, were slow to react. Many preferred the relatively safe haven of ice age dynamics to the difficult and politically noisy realm of contemporary environmental science. Broecker's bold hypothesis seemed almost apocalyptic, a rash and untested scenario that added a new level of uncertainty and controversy to a subject that already was too uncertain and controversial. Climate science's ambition to predict the future was going to be more difficult to achieve than many supposed, he warned, and perhaps decades farther in the future. More than that, the future itself was likely to be more dangerous. For computer modelers and others looking for a baseline of natural variation to simulate the impact of the industrial age on climate, the physics of gradual change was going to be difficult enough to absorb. What Broecker offered was something even more intractable—a more lethal and more willful system controlled by the tangled dynamics of chaos. About all that Broecker promised was more years of hard work and, at the end of this longer and bumpier road, a future of greater uncertainty.

In a 1987 commentary in the journal *Nature*, Broecker wrote that as they contemplated the future, climate scientists had been "lulled into complacency by model simulations that suggest a gradual warming over a period of 100 years" and by the oxygen isotope record in deep-sea cores that gave the impression that the response of the climate system to changes wrought by subtle alterations of Earth's orbit of the Sun is smooth and gradual. "Only recently have we begun to realize that this impression is a false one," he wrote. Looking back, clues were there in the North Atlantic seabed in the changing species of planktonic organisms that probably reflected sudden changes in sea surface temperatures during the past 100,000 years. "It took more than this, however, to make us take these abrupt changes seriously," Broecker confessed. "The evidence that turned our heads came from holes drilled through the Greenland ice cap."

Employing an analogy he would use more than once, Broecker wrote: "We play Russian roulette with climate, hoping that the future will hold no unpleasant surprises. No one knows what lies in the active chamber of the gun, but I am less optimistic about its contents than many."

What the world needed was a cadre of young scientists dedicated to intensifying investigations of a range of climate issues, including a more careful examination of the evidence of change in ocean sediments and ice cores.

"Although we don't know nearly enough about the operation of the Earth's climate to make reliable predictions of the consequences of the build-up of greenhouse gases, we do know enough to say that the effects are potentially quite serious," he warned. A climate that experienced sudden leaps could deal wildlife a serious blow. Our food supply might be at risk. "To date, we have dealt with this problem as if its effects would come in the distant future and so gradually that we could easily cope with them. This is certainly a possibility," he wrote, "but I believe that there is an equal possibility that they will arrive suddenly and dramatically."

For all its dark warnings and pessimism, Broecker's message struck a responsive chord among young graduate students entering the earth sciences. Here was an uncrowded field, a virtually new science with a new sense of consequence and urgency to embrace. As more and more young scientists turned to the Greenland ice core records and sought to develop their own lines of research, a problem that had been simmering for years quickly became more serious. There was not enough Greenland ice to go around. As more scientists began to focus on the climate profile from the ice sheet, the origins of the cores became a more serious issue. None of the deep-core ice had been drilled from scientifically optimum locations on the ice sheet.

As a young and essentially unproven line of research, ice core drilling in Greenland had always been forced to compromise with logistical convenience. The enterprise, cumbersome and costly, had benefited for years from the willingness of the U.S. Army's Corps of Engineers to explore the ice as part of its strategic interest in Greenland. But that support had come at a price. In the 1960s, the first core to bedrock had been drilled in northwestern Greenland at a site that was chosen primarily because of the nearby location of Camp Century, the "City Under the Ice," the military's elaborate experiment with habitation in the ice cap. In the 1970s, after military interest faded, Langway, Dansgaard, and Oeschger, who had searched the ice sheet for ideal sites for drilling to bedrock, presented the National Science Foundation (NSF) with specific ideas about extracting a core from a site chosen for maximum scientific benefit—the summit of the ice sheet. Again, however, even as the U.S. civilian science funding administrators committed $10 million to a second surface-to-bedrock core, the scientists of the Greenland Ice Sheet Program were required to settle for a much inferior southeastern drill site, far from the summit, because it was logistically convenient to a Distant Early Warning radar station known as Dye 3.

By the late 1980s, the scientific landscape was very different. "We've got people tripping over each other to do this research," Herman Zimmerman, an NSF program manager, told journalist Elizabeth Pennisi in 1989. So intense was the international competition for ice samples and research participation in a drilling program in Greenland that Broecker, Dansgaard, and Oeschger concluded a bold and ambitious agreement. Two separate ice cores would be drilled at Summit, just 20 miles from one another. The NSF would finance GISP2, a $25 million multidisciplinary project involving researchers from 12 U.S. universities. For a like amount, a European consortium would undertake the Greenland Ice Core Project, GRIP. The great northern ice cap was about to reveal its secrets. And the science of climate change would never be the same.

8

MOMENTS OF TRUTH

"Large, rapid, and global" change in Summit ice

For one team of young U.S. researchers working on the Greenland Ice Sheet Project II, almost all of them new to the science of polar ice drilling, the moment they had been waiting for arrived in the middle of the summer of 1992. After more than five weeks, members of the crew had acclimated themselves to the conditions at Summit and were coordinating their tasks about as well as could be expected of any group at an elevation of 10,400 feet in subzero cold. They were working in a glorified snow cave known as "the science trench," field-testing and processing the 5.2-inch core for later laboratory examinations that would subject polar ice to a multifarious array of chemical and physical analyses.

At the front of the processing line, Kendrick C. Taylor, a young geologist from the Desert Research Institute in Reno, Nevada, was running a direct electrical current through the core. This ingenious electrical conductivity measurement (ECM) technique had been invented by Claus Hammer, a Danish geophysicist and an experienced member of the team of European scientists who were drilling a twin ice core just 20 miles east of the U.S. site. Kendrick applied electrodes to the side of the core that had been shaved smooth. He watched as the acidity variations of the ice changed the strength of the current and made green wiggles on the screen of his computer.

Every inch of the two-mile core would be tested for its electrical conductivity, and a continuous profile of the ice sheet would be assembled. Out of the green wiggles came a pattern of annual layers, as small but regular variations tracked subtle seasonal differences in the ice's acidity. A spike in the readings might signal the presence of aerosols from a volcanic eruption somewhere in the Northern Hemisphere. A dip might indicate the fallout of airborne soot wafting across the ice sheet from distant forest fires. Taylor had connected the device to a speaker for a time, giving the researchers an audible sense of the variations in the signal's strength as the electrodes moved down the core. This first-pass procedure didn't identify the causes of these spikes, but it quickly generated a continuous profile that located interesting places for other researchers to look.

In line behind Taylor was Richard B. Alley, a young geoscientist who was applying the oldest test in the books. Following in the footsteps of the pioneers who had first examined polar ice, Alley was looking through the ice and seeing what he could see in the diffused rays of light emanating from a fluorescent bulb on the other side of the core. In the literature, this critically valuable albeit imperfect procedure has been dressed up in Sunday clothes and given the name of "visual stratigraphy." Like anything so entirely human, however, it really is as much art as science, demanding personal patience and practice and defying attempts to objectively replicate its results photographically or archive its data electronically.

Alley was looking for anything interesting that might show up, of course, but first and foremost he was counting back the age of the ice by identifying the individual layers along the core. He passed the core sections in front of fluorescent lights that showed the variations in lightness and darkness caused by differences in physical properties such as trapped air bubbles or lines of dust.

Besides leaving regular variations in electrical conductivity, the Arctic seasons' dramatic differences, summer's perpetual sunlight and winter's perpetual darkness, left behind a telltale visual signature in the ice. Summer snowfall was coarser than winter snowfall, and this difference in texture was preserved as a subtle layering effect as snow turned to ice, allowing Alley's practiced eyes to discern the passage from one year to the next.

This rookie Penn State professor had looked at more polar ice than most Americans in the science trenches at the time, having spent two field seasons in Antarctica working on a Ph.D. thesis on how snow turns to ice. In fact, Alley was one of the few young American scientists at GISP2 with any prior experience in

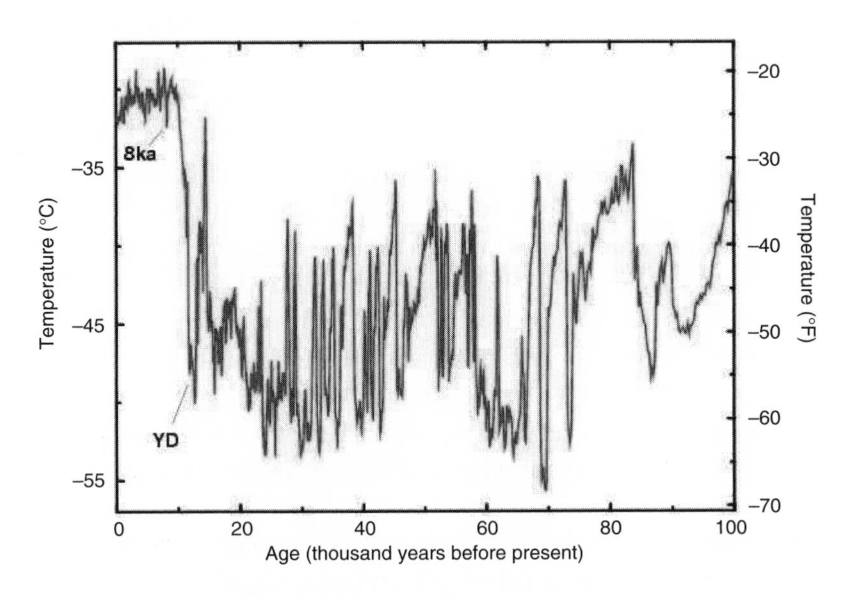

Temperatures in Central Greenland

This profile of temperatures in central Greenland during the past 100,000 years, reconstructed from oxygen isotope measurements along the GISP2 ice core in the mid-1990s, illustrates a climate pattern that is very different than the slumbering stability that researchers traditionally envisioned for ice ages. Notice the Younger Dryas, the last abrupt cold period before the current warm era, and the brief "cold snap" 8,000 years ago. Reprinted from National Research Council, *Abrupt Climate Change: Inevitable Surprises*, National Academy Press (2002).

Greenland ice. In the summer of 1985, as a graduate student, Alley had been one of a small U.S. contingent that helped a team of Danish experts survey the ice cap as part of the process of selecting the best site for what eventually became the European GRIP and U.S. GISP2 projects.

As almost everyone had expected of the Summit of Greenland, the ice under the site already had proven its value as an ancient climate archive, delivering up continuous sections that were beautiful, as polar ice cores go, and superior to the cores from Dye 3 and Camp Century. Far from the divide and the principal accumulation area of the glacier, the annual layers in the shorter cores taken from the earlier drilling sites were subject to more distorting effects of ice flow. At the twin Summit drilling sites, scientists were able to easily observe eras of climate just halfway down through the ice sheet that in the earlier cores were razor-thin layers near bedrock. By the middle of the summer of 1992, the Americans had drilled through a mile of ice, although they were still waiting to encounter what they had spent all of this money and time on and come all this way to see.

Where was the ice that would prove—or disprove—the truth of abrupt climate change? For more than a decade, the idea that climate could change noticeably within a single human lifetime had been bandied about among climate scientists but taken seriously by only a few. Not many researchers had actually seen the data from Camp Century and Dye 3, which had been the subject of long discussions about what, among many, circumstances could make them faulty. In any case, even those who were intimately familiar with the evidence would concede that episodes of abrupt climate change were "only suggested rather than proven" by the earlier cores. So where was the section of core that would settle the arguments once and for all?

The Greenland ice sheet concealed the secret of rapid climate change almost perfectly. This critical layer of ice, just a few feet thick, was suspended almost exactly in the center of the ice sheet at Summit—about a mile below the ice sheet's surface and a mile above bedrock. Drilling back in time, there were really no signs of it coming. The ice they were studying at the time looked a lot like the ice that had fallen as snow during the last 10,000 years. Yet they knew they were close.

They knew they had the critical layer of ice in their snow cave. Wanda Kapsner, a Penn State graduate student, had been taking thin sections about every 20 meters along the lengths of core laid out in the cave. She told Alley, "This section is in Holocene ice and the next section 20 meters down is in Ice Age ice, and so between these two is where you're going to find it."

This team of scientists was about to complete their six-week stint at Summit, and a new team was about to take its place. So it was up to project leader and paleoclimatologist Paul Mayewski to decide which team was going to handle this important ice. On hearing the news from Kapsner, Mayewski told Alley, "Fine, before we get out of here, we're going to do that ice."

The ice that had formed from falling snow during the transition from the last of the cold, dry, windy ice ages to the first of the warm, wet calms of the modern 10,000-year-long Holocene climate is 1,678 meters, just over a mile, down the GISP2 core. Rendered in ice, what exactly would it look like, this boundary of epochs? The young American scientists had read the literature from Chet Langway, Willi Dansgaard, Hans Oeschger, Wally Broecker, and others, and they had heard from the Europeans, who were about a year ahead of them in drilling at Summit. Yet still they were not entirely prepared for what they saw that day in the ice, for the suddenness of it.

"You did not need to be a trained ice core observer to see this," recalled Alley. "Ken Taylor is sitting there with the ECM and he's running along and his green line is going wee, wee, wee, wee—Boing! Weep! Woop! And then it stays down." Dust in the windy ice age atmosphere lowered the acidity of the core to a completely new state. "We're just standing there and he just draws a picture of it," Alley said.

Spontaneous celebration was followed by a sudden and unexpected quiet. "I think we cheered," recalled Alley, "and then we were all a little sobered. Because it was just so spectacular. It was what we'd been looking for, and there it was, and then we're sitting there. Holy crap."

The instant of recognition that summer of 1992 had a raw feel to it, although eventually the disquiet would find concrete expression in numerous articles and presentations as the scientists became accustomed to the large truth of abrupt climate change and immersed themselves in its fine details. Alley recalled later: "Those of us who were down there in that trench at that time knew right then that our picture of the world had changed. There's a whole bunch of us who came out of that ice core project who have since dedicated ourselves to understanding abrupt climate change."

"My attitude changed profoundly," Kendrick recalled five years later in an article for *American Scientist* magazine. Before that experience in Greenland, he wrote, "I used to believe that changes in climate happened slowly and would never affect me."

Reviewing the evidence from Greenland and later results from ocean sediments, Taylor observed that in a span of just 20 years many regions of the world had experienced the changes he had first seen in the ice. "There was no warning," Taylor wrote. "A threshold was crossed, and the climate of much of the world shifted abruptly from cold to warm. This was not a small perturbation; our civilization has never experienced a climate change of this magnitude or speed."

Later, archaeologists and anthropologists would give human dimension to the evidence of change in that section of Greenland ice core. On one side were the atmospheric remnants of the climate of the Younger Dryas, a cold and windy, dry and unstable regime that had driven the Neolithic human clans to turn to agriculture. On the other side was the warm and wet and calm climate of the Holocene that during the past 11,000 years had cradled the rise of civilization. What Taylor and Alley saw was a boundary that looked nothing like the subtle progression of blurring that might have been left by a congenial process of gradual climate change. Holy crap. The line in the ice, just a few inches thick, looked more like a trap door falling shut.

In the GISP2 science trench, the tray holding the section of core rolled down the assembly line and then it was Alley's turn at the ice. "It slides across in front of me and I'm trying to identify years: 'That's a year, that's a year and that's a year, and—woops, that one's only half as thick.' And it's sitting there just looking at you. And there's a huge change in the appearance of the ice, it goes from being clear to being not clear, having a lot of dust."

Long before chemical analyses would be conducted in more commodious laboratory environments in Europe and the United States, the scientists in the snow cave that day knew what this relic of ancient atmospheric processes would mean in terms of the science of climate. Reduced to a matter of inches, along a section of ice that went from clear to opaque, was environmental change of singular power and scope. "We knew what you'd see very clearly with an isotope record or a chemical record," said Alley. "We knew just standing there looking at it that it was huge, it was very fast, it did this bounce. . . . "

The bounce was part of climate change on a timescale that had never been seen before. In the summer of 1992, the quality of the GISP2 core and the sensitivity of the electrical conductivity measurements made it possible for Taylor to obtain more than 15 samples for every annual layer in the ice that had fallen as snow more than 10,000 years ago. In one sense, the bounce was part of a pattern of discovery that historians of science would recognize. Every time researchers had found a new way to a more sharply focused view of climates past, episodes of change emerged that had not been seen before. With each more finely detailed rendering, the climate system had emerged as more varied and more precariously balanced.

The following February, in *Nature*, Taylor described a "flickering" pattern between colder and milder conditions that lasted 10 or 20 years and typically terminated in less than 5 years. "The transitions between century or longer warm and cold periods are characterized by abrupt fluctuations in alkaline dust concentrations that occur over periods of less than 10 years," Taylor wrote. During the warm Dansgaard-Oeschger events of the ice age and "at main climate transitions, the part of the climate system influencing the ECM is behaving like a flickering switch that fluctuates between two states before stabilizing."

A "reordering of atmospheric circulation" seemed the most likely agent for such rapid changes, especially a change in the speed of winds bearing calcium carbonate dust over the ice sheet. Changes in atmospheric circulation also seemed the likely cause of dramatic change in another climate parameter. In the summer of 1992, this one stuck out of the ice core when U.S. researchers got their first look at the seam between the Younger Dryas and modern climate.

Alley's eyeballing and Taylor's ECM readings agreed: The changes in thickness of the individual layers in the ice were unmistakable. It may have been the fastest climate change ever seen. In April 1993, in a brief article in *Nature*, Alley, Kendrick, and their team reported that during the transition from the Younger Dryas to the Holocene, the amount of snow falling over the summit of Greenland had doubled, and the change seemed "to have occurred in three years with most of the change in one year." The researchers estimated that the sharp change in layer thickness at 1,678 meters, about a city block more than a mile deep in the ice sheet, corresponded to 11,640 years ago.

Again, the speed of events seemed to rule out the more slowly changing elements of the climate system such as solar radiation, atmospheric CO_2 changes, or glacier retreat. "Causes must be sought in threshold levels in the climate system or in especially fast-acting components of the climate system including restricted, sensitive 'trigger' regions of deep oceanic convection, and atmospheric circulation patterns," Alley wrote.

Whether it jumped or was pushed, there was no doubt that the atmosphere had changed over the North Atlantic. On July 9, 1993, Mayewski and several members of the team reported in *Science* that numerous chemical analyses of the Younger Dryas ice "provide evidence of an extremely dynamic atmosphere thus far unparalleled in the Holocene. Investigation of the forcing of such an extreme environmental state offers a new view of climate change." Temperatures had warmed during the transition, of course, although a study by Wanda Kapsner in 1995 showed that rising temperatures alone could not account for the doubling of snowfall. The cold polar front had retreated northward, and the storm track was more squarely over Greenland.

In 1997, after more detailed laboratory work, Taylor reported in *Science* that changes at lower latitudes had taken place perhaps 15 years before the last jump to the Holocene in Greenland. Climate proxies related to areas outside the Arctic apparently changed before proxies of Arctic climate. Taylor said he couldn't tell "if the climate transition started at lower latitudes or if earlier changes that we are unable to detect occurred in more northern regions."

Chemical analyses of ice that had registered a spike in electrical conductivity measurements pointed to the eruption of a volcano that either was nearby in Alaska or Iceland or was a large explosive event elsewhere, 11,660 years ago. "We do not believe the [approximately] 10-year cooling associated with this eruption would have been sufficient to tip a precariously poised climate system to a warmer state," Taylor wrote, "but the coincident occurrence of the eruption and climate transition is intriguing." In ice representing the following 15 years, other chemical evidence pointed to a decrease in dust, indicating either an increase in moisture in areas outside the Arctic where the dust originated or a general drop in wind speeds.

A full field season ahead of the Americans, 40 members of a European team that included such polar ice pioneers as Willi Dansgaard and Hans Oeschger reached into silty ice at the bottom of the ice sheet at Summit in the summer of 1992. In contrast to the Americans, who first focused on the data of the Younger Dryas, the Europeans took a more comprehensive theoretical approach to their climate profile and its implications for modern conditions.

Taking in this first view of the climate profile from top to bottom, they came across features that were entirely familiar to Dansgaard and Oeschger. Echoing their earlier findings, the sudden spikes of warmth depicted a climate shifting "between two apparently quasi-stationary climate stages." It must have been a deeply satisfying recognition. The 23 warm "interstadials," or

"Dansgaard-Oeschger events," that first appeared 25 years earlier in the oxygen isotope analyses of Camp Century ice were just where the Europeans expected to find them along their Summit core.

Reporting their first results in *Nature* that September, Sigfus Johnsen declared an end to the debate over the reality of this long, irregular series of abrupt climate changes that punctuated the last ice age. The common argument—that they were figments of Camp Century and Dye 3 records that had been skewed by deformation of ice layers near bedrock—found no support in the superb climate archive at Summit. "The results reproduce the previous findings to such a degree that the existence of the interstadial episodes can no longer be in doubt," Johnsen wrote. Rapidly, time and again, temperatures had climbed about 12.6°F, although still 9°F below modern levels, and in irregular steps over the next 500 to 2,000 years returned to ice age cold.

Johnsen raised a point that would haunt other researchers. If the abrupt changes turn out to reflect "a randomness" in the climate system, if two or more patterns are possible from a given set of variables, "climate modelers will have to reconsider the feasibility of predicting natural climate change."

Computer modelers would struggle for years with the problem that Johnsen described as randomness. Their most powerful simulations are highly complex mathematical creations that capture many processes. These numerical weather prediction models are the best things that have ever happened to weather forecasting, but they have their limits. On the timescale of weather, the atmosphere exhibits a kind of chaotic behavior that typically causes the skill of even the best forecasts to deteriorate after a few days. Johnsen's speculation about the origin of the abrupt-change "interstadials" was not good news to researchers who had higher hopes for the predictability of climate. The idea of "two or more possible flow schemes for a given set of primary parameters" reminded everyone of the kind of chaotic nonlinear behavior that imposes limits on weather forecasting.

Modelers, who might take refuge in the idea that warm climates behave differently than ice ages, weren't going to find much encouragement from Johnsen either. The last millennium had seen a medieval warm period in Europe that was followed by a "little ice age" during which sea ice frequently surrounded Iceland. The cold temperatures had been succeeded by a warming that culminated in the 1920s with an abrupt rise in temperatures—"too abrupt to be explained by the increasing greenhouse effect." These oscillations had been smaller than ice age changes, to be sure, but followed the same "sequence of events (gradual cooling followed by abrupt warming)."

The Europeans were first to examine a section of ice far down their Summit core that the leaders of both projects considered a prime target for research. Below the depth of 2,788 meters (1.7 miles),they found ice that had fallen as snow the last time Earth's climate was as warm as the modern Holocene. The "interglacial" warm era known as the Eemian began about 135,000 years ago and ended with the beginning of the last ice age 110,000 years ago. Because other paleoclimate proxy records suggested a time of relative stability when global temperatures were somewhat warmer than during the Holocene, the Eemian period had long been considered an important "analogue" for the future of modern climate under the influence of an intensifying greenhouse effect. The findings from the European core, announced in *Nature* in the summer of 1993, surprised everyone.

The same episodes of rapid and large change that characterized the last ice age ran straight through the Eemian period and into the earlier ice age. Down in the core below 2,750 meters, in ice that the Europeans were confident represented the period of Eemian warmth about 120,000 years ago, oxygen isotope data showed two especially large and sudden plunges toward ice age cold. In one episode, average temperatures apparently plunged 25°F for about 70 years. The only period of relative stability during the Eemian came during the last 2,000 years of its warmest stage.

"The unexpected finding that the remainder of the Eemian period was interrupted by a series of oscillations, apparently reflecting reversals to a 'mid-glacial' climate, is extremely difficult to explain," the Europeans wrote. "Perhaps the most pressing question is why similar oscillations do not persist today, as the Eemian period is often considered an analogue for a world slightly warmer than today's." Given the history of the last 150,000 years, they wrote, the past 8,000 years "has been strangely stable."

In an accompanying commentary, American climatologist James W. C. White, a stable-isotope specialist, asked if the Eemian record in the Europeans' ice core was a glimpse of our own future. "Whatever the answer to that question, the speed with which the climate system can shift states gives us pause. Adaptation—the peaceful shifting of food growing areas, coastal populations and so on—seemed possible, if difficult, when abrupt change meant a few degrees in a century. It now seems a much more formidable task, requiring global cooperation with swift recognition and response."

As journalist Richard A. Kerr noted in *Science*, the news from Greenland shattered the conventional view among climatologists of the warm intervals between ice ages as "benign interludes." On the front page of the *New York Times*, science writer Walter Sullivan reported: "To the astonishment of climate specialists, an analysis of ice extracted from the full depth of the Greenland ice sheet has shown that except for the 8,000 to 10,000 years since the last glacial epoch, the climate over the past 250,000 years has changed frequently and abruptly." Michael Morrison, associate director of the U.S. project, where bedrock had been reached just two weeks earlier, told Sullivan that the Americans expected to confirm the European findings.

It was not until U.S. researchers got a close look at the deep sections of their core that they noticed serious discrepancies between the new long climate profiles. The U.S. core gave the same general impression of instability during the Eemian warm period, but when researchers examined the data in detail, they found no real agreement in the sequence of individual climate episodes. From the surface all the way down to 2,780 meters, through 90 percent of the depth of the ice sheet, the two finely resolved climate variation profiles agreed with one another in almost lock-step precision. Then, unexpectedly, at a point that was still 200 meters above bedrock, agreement broke down completely.

In December, two reports in *Nature* broke the unhappy news. Taylor for the U.S. team and Claus Hammer for the Europeans coauthored a report that described major discrepancies in their electrical conductivity measurement data for the ice cores at the depths below 2,750 meters. And a paper by Pieter M. Grootes, White, Johnsen, and others reported similar problems with the oxygen isotope readings at the Eemian depths. Everyone was confident that the upper 90 percent, where both cores agreed, represented two highly detailed, accurate records of climate going back 110,000 years that were unaffected by ice flow. However, with the integrity of the last 10 percent of at least one ice core, and possibly both, in doubt, the Europeans' interpretation of their Eemian results was in serious trouble. Arguing that evidence of disturbance seemed more apparent in the Americans' core, several members of the European team held out for the integrity of their own data back through the Eemian.

Yet just as the conformity of the two climate profiles proved the reality of abrupt change for the last 110,000 years, their nonconformity proved critically valuable in detecting faulty data in older ice. The circumstances threw a new light on the seemingly extravagant decision to simultaneously undertake two large and nearly identical ice coring projects in central Greenland. Without a twin of the same ice, the ability to quickly test one against the other, climate science might have been led down a long and unrewarding road.

Later studies finally persuaded scientists on both projects that ice flow had distorted the sequence of the annual layers in both cores below about 2,800 meters. Strong evidence came in 1995 in a study by Alley and others of the physical properties of the deep sections of both cores. Most convincing was an analysis by the French glaciologist Jérôme Chappellaz of the atmospheric gases trapped in the air bubbles. Chappellaz compared concentrations of methane in the supposed Eemian-age Greenland cores with bubbles of the same age in ice from the Vostok core in Antarctica. Because methane, a climate "proxy" for the extent of continental wetlands, is globally mixed in the atmosphere, the Greenland and Antarctic data should agree in general direction even if the amplitude of the signal is different. By 1997, the issue was no longer in doubt: Where both of the Greenland cores showed large and rapid variations in methane trapped in their deep ice bubbles, Chappellaz reported, "the Vostok record shows no such variations."

In an effort to overcome the "Eemian problem" in the ice, to extend the sharply focused view afforded by polar ice through the last period of warm climate, the Americans and Europeans joined in 1996 to finance the drilling of a third central Greenland ice core. Drilling at a promising location north of the Summit site, the North Greenland Ice Core Project proved even more frustrating. After seven years of stuck drills and other stops and starts, in the summer of 2003 NorthGRIP drillers finally reached bedrock at a depth of 3,085 meters. There, unexpectedly, they found that heat rising from some unknown geothermal process had melted the Eemian ice.

In the end, of course, whatever the ice may have revealed of the Eemian period, leaders of both projects recognized that they had rewritten the history of climate over the last 110,000 years. Scores of international scientists had analyzed polar ice as never before, producing two long, reliable, and continuous profiles of ancient climate with a clarity that was unprecedented. No longer was there any doubt about the warm Dansgaard-Oeschger events. Twenty-four episodes stood out clearly, as did 13 separate Heinrich events that dramatically plunged the Northern Hemisphere into especially severe cold and dispatched armadas of icebergs across the northern Atlantic. Scientists would spend years poring over the results, comparing them with other climate archives, looking for ways to explain this warming or that chill.

More interesting to the larger community of climate scientists—and to nonscientists—was the unmistakable pattern that emerged. This was not the record of ancient climate that anyone had been taught to expect. It was not just that the climate was subject to more variation than was generally supposed, although the idea of so much change during an ice age was certainly a surprise. More surprising, of course, and most interesting was not the fact of change but its suddenness.

"From the central Greenland ice cores we now know that the Earth has experienced large, rapid, regional to global climate oscillations through most of the last 110,000 years on a scale that human agricultural and industrial activities have not yet faced," an international group of leading climate scientists wrote in a special issue of the *Journal of Geophysical Research* devoted to detailed studies of both cores. The work reported was part of a flood of new information that poured out into a growing community of climate scientists who were increasingly attuned to changes taking place in the atmosphere and around the world. The Dane Claus Hammer, the American Paul Mayewski, David Peel of the British Antarctic Survey, and Dutch-born Minze Stuiver wrote: "The ice-core records tell a clear story: humans have come of age agriculturally and industrially in the most stable climate regime of the last 110,000 years. However, even this relatively stable period is marked by change. Change—large, rapid, and global—is more characteristic of the Earth's climate than is stasis. Until we understand the operative mechanisms, it will not be possible to understand current change or predict future change."

So the secret of the Greenland ice was out. No longer was abrupt climate change the hypothesis of a small group on the strength of debatable data. The evidence was unusually reliable and voluminous, taken from twin sites analyzed by two large, well-organized research teams from more than 40 university and national laboratories. As Taylor described the collaboration, "We shared samples, spent time in one another's labs, replicated one another's results, proposed ideas, tore them apart and then jointly proposed better ones."

Many scientists encountered abrupt climate change for the first time when they read the papers by the Europeans and Americans in the scientific journals in 1993. In December, the Greenland scientists made joint presentations and gave interviews during the fall meeting of the American Geophysical Union in San Francisco. Reporting on the meeting, many newspapers around the world published their first articles on the subject.

Among climate researchers, the news from Greenland encouraged a range of new studies. Researchers like Taylor and Alley would go looking for longer high-resolution records of climate history in Antarctic ice. Oceanographers and other investigators would search out and find more evidence of abrupt change in ocean sediments and a variety of other climate archives around the world.

The remarkable successes of the European and U.S. ice core projects at Summit are commonly seen as occasions of original discovery. Scientists familiar with the paleoclimatology literature would see them more accurately as verification of the earlier work in the 1960s and 1970s by Dansgaard, Oeschger, and Langway. Perhaps it is in the nature of such "Big Science" projects, like glaciers moving down a mountain, to obliterate the records of earlier exploits with their massive progress. Science, after all, is not a museum of ideas, but a process of finding out how the world works. You test ideas discard what doesn't seem to work, and move on. Scientists have a keen sense of their debt to those investigators whose ideas and data they employ, but they don't necessarily have a keen sense of history—even the history of their own discipline. So perhaps it was only natural that no one in the mid-1990s seemed to see the success of the Greenland ice core projects as the culmination of investigations that began 63 years earlier in a very lonely place called Eismitte.

What would it mean to Ernst Sorge, the German researcher who dug the first trench down from his cave in the ice cap that winter of 1929? Or to the expedition's organizer Alfred Wegener, the father of continental drift, who died on the ice that winter—would he recognize this new paradigm in climate science?

The passage of 63 years can be a long time in a rich and actively pursued science, of course—although, as almost everybody knows, it's really no time at all on the geological scale. Who would have thought that 63 years is time enough for climate to change?

9

SEDIMENTATIONS

Finding abrupt change around the world

Confined to the remoteness of Greenland and the special techniques of polar ice analysis, the meaning of the early record of abrupt climate change might never have been widely recognized. For a decade and more after publication of the Dye 3 results, the Greenland researchers' idea of rapid, radical change on timescales that could be measured in decades or even individual years attracted very little interest. Even scientists willing to accept the validity of the data were inclined to a more conservative interpretation: They must have been regional. In the 1980s, who knew what peculiarities of Arctic weather might have caused those strange wiggles in the climate profile of Dye 3? And if it were really so rapid, it just must be the exception that proved the rule. As Darwin himself had proclaimed: *Natura non facit saltum*—Nature does not make leaps. The idea was just too far outside science's basic teachings to be taken very seriously. Moreover, no other archive of ancient climate yielded the kind of detail that the Greenland researchers claimed to see in polar ice.

In the mid-1990s, the results of the Summit projects emerged into a different scientific landscape. As signs of climate change were becoming apparent, greater future change was an increasingly potent issue, and so was the need to understand the pattern and pace of climate's natural variations. Suddenly, from the Summit projects came this incredible, high-fidelity continuous profile of climate's behavior through the last 100,000 years. For the quality of analysis, for their exhaustive detail, the projects set new standards for such investigations and just as critically important was that by the mid-1990s, signs of abrupt change were showing up in other archives beyond Greenland.

In tandem with the ice core findings were important discoveries by oceanographers, the one group of researchers that had been both well positioned and temperamentally inclined to take on the challenges posed by the early Greenland findings. The advent of postwar nuclear chemistry and advances in marine engineering had given the old science a young face and a reputation for overturning conventional geological thinking. It was oceanographers whose discovery in the 1950s of spreading midocean ridges that had sparked the theory of plate tectonics and finally confirmed Alfred Wegener's concept of continental drift.

Outside of Greenland, perhaps the most important center of activity was the Lamont-Doherty Earth Observatory in New York, where theorist Wally Broecker and others had been on the trail of abrupt change since the mid-1980s. Moreover, Lamont possessed its own treasure trove of paleoclimate data waiting to be tapped. For years, scientists using the facility's research vessels had labored under a standing order from Maurice Ewing, Lamont's founder and first director, to extract a sediment core every day they were at sea. Beginning in the late 1980s, even as the big ice core drilling projects were just getting under way at Summit, an unlikely researcher using Lamont's enormous depository of ocean sediment cores was opening a new line of research into abrupt climate change.

Gerard C. Bond was not an oceanographer or a climatologist and had no particular interest in the science of abrupt change. Years later, accepting the prestigious Maurice Ewing Medal of the American Geophysical Union (AGU) for this work, Bond would describe himself as a relative newcomer to paleoceanography who had spent half his career as a "real geologist," more interested in the composition of rocks—petrology—and research projects far removed from the ocean. The sharp turn of events grew out of a research grant proposal that Bond drafted with a colleague, geologist Michelle A. Kominz, who had developed a new optical analysis technique. Bond and Kominz proposed to test the new method on the color record of a North Atlantic deep-sea core from the Lamont library, to see if it identified the large-scale orbital cycles in Cambrian Age rocks, and the proposal went to Broecker for review.

"Wally rushed into my office telling me that the core's color record instead revealed the long-sought marine imprint of Greenland's Dansgaard/Oeschger cycles," Bond recalled at the AGU award presentation in San Francisco in 2003. "I had never heard of Dansgaard/Oeschger cycles, but Wally and other Lamont paleoclimatologists . . . managed to convince me that they were much more interesting than Cambrian cycles. By the early 1990s, I had shifted my research from rocks to deep-sea mud. In this new field I was surrounded by a baffling array of machines with flashing red lights, toxic chemicals, and coworkers who spoke the languages of chemists, physical oceanographers, and modelers."

Initially, at least, the change in fields was not as great a leap as he might have expected. The North Atlantic deep-sea cores that Bond examined contained grains of material dropped by melting icebergs that were much the same as the sedimentary rocks that were his first geological specialty. He found himself back in the business of petrology, microscopically studying rocky grains in the seabed to determine their origin. Bond showed how far and wide the iceberg armadas of the Heinrich events spread across the North Atlantic and how their timing matched the cold phases of the Dansgaard-Oeschger cycles in the Greenland ice.

Willi Dansgaard quickly responded to the news from Lamont-Doherty by incorporating the North Atlantic findings into the latest Greenland data. So it happened that Bond was the single American coauthor of the famous 1993 *Nature* article in which Dansgaard and other members of the European Greenland Ice

Core Project team first described the results of their project at Summit. Alongside the new oxygen isotope profile from Greenland was a depiction of the visual color record from the Lamont Deep Sea Drilling Program sediment core 609 that had provoked Broecker's excitement and Bond's conversion to paleoceanography. In Dansgaard's estimation, the seafloor record was comparable to the ice core profile. Except where grains of ice-rafted carbonate "corrupt the grey scale," Dansgaard matched nearly all of the oxygen isotope shifts in the Greenland ice with the color record in Bond's sediment core.

As it happened, the white layers composed of carbonate grains in the core that Dansgaard called corruptions of the color record were features that led Bond to a dramatic advance in the science of abrupt climate change. Using his experience in studying the composition of rocks, he identified the source of the glacially scoured rocks in the debris layers and the icebergs that carried them across the Atlantic. Bond demonstrated that the ice age climate was more complex and even more extreme than the profile drawn from the ice cores. Not only was the last ice age punctuated by abrupt episodes of warming that lasted 1,000 years or more, but the seafloor sediments revealed other instances when temperatures rapidly plunged to levels of exceptional cold.

In 1992, Bond and an international team that included Broecker and the German marine geologist Hartmut Heinrich confirmed Heinrich's identification of a type of abrupt climate change in the North Atlantic that seemed to be very different from the events that dominated the Greenland record. In 1988, Heinrich had described telltale layers of ice-rafted debris in sediment cores taken from an area of the northeastern Atlantic known as the Dreizack Seamount. Bond combined the Dreizack data with Lamont data from cores extracted from across the North Atlantic containing microscopic rock particles that fell to the ocean floor during six different "short-lived, massive discharges of icebergs originating in eastern Canada" between 14,000 and 70,000 years ago. At the same times, temperatures had fallen in the atmosphere over Greenland and on the surface of the ocean over a vast area of the North Atlantic.

"The cause of these extreme events is puzzling," they wrote in the journal *Nature.* "They may reflect repeated rapid advances of the Laurentide ice sheet, perhaps associated with reductions in air temperatures, yet temperature records from Greenland ice cores appear to exhibit only a weak corresponding signal."

Bond went on to identify a large cycle of warm periods and progressively more extreme cold periods that ran through the last ice age at intervals of 10,000 to 15,000 years. In 1993, in a *Nature* article that linked ice core and ocean sediment records, Bond and colleagues described a sawtooth pattern that was composed of a succession of warm-cold Dansgaard-Oeschger cycles that were "bundled" between Heinrich event cold extremes. The pattern would come to be called "Bond cycles."

"The series of saw-tooth shaped cooling cycles is clearly a fundamental structure of the atmosphere and sea-surface records, and must bear a close relation to the Heinrich events and the repeated, massive collapses of the Laurentide ice sheets," they wrote. For the first time, Bond had documented a link between the ice sheets, the ocean, and the atmosphere. "With the evidence in hand we cannot be certain whether the cooling cycles were caused entirely by internal oscillations of the ice sheet, or whether they reflect a mode of climate forcing that caused ice sheets to grow, culminating each time in a prolonged, cold stadial, ice-sheet instability and massive calving." In other words, they could not identify the mechanisms at work. Was the ice sheet driving the changes in climate? Or was some other climate process causing the ice sheet to become unstable periodically?

As Bond looked more deeply into the layers of debris, the mystery grew more complex. In 1995, Bond and Rusty Lotti, curator of Lamont's Ocean Sediment Core Laboratory, found traces of dark basaltic glass in the layers that could only have been scraped from the volcanoes of Iceland by glaciers. This evidence that more than one ice sheet was involved caused them to question the idea that the dynamics of the Laurentide ice sheet were responsible for Heinrich events. Perhaps other changes were causing all North Atlantic ice sheets to advance and retreat.

Whatever their cause, the "bestiary" of abrupt climate changes during the ice age was becoming larger and more interesting. There were Dansgaard-Oeschger events—episodes lasting 1,000 years or so when temperatures rapidly warmed, slowly staggered back over several centuries and then suddenly plunged to ice age cold. There were Heinrich events—times of exceptional cold when armadas of icebergs drifted far and wide over the North Atlantic. There were Bond cycles, bundles of warm-cold D-O events culminating in a frigid Heinrich event. The mechanisms that caused these episodes would be the subject of years of debate and contemplation and, beginning in the 1990s, the objects of a wide-ranging effort to discover their global reach and the abrupt upheavals in climate they implied.

While Bond and colleagues were able to show that the sediments in the North Atlantic bear the imprint of changes noticed in Greenland's ice, the rapid pace of these events could only be inferred from the sharpness of the Heinrich layer contours. The detail in the ocean sediment could not match the resolution of change in polar ice. In most areas of the open ocean, the sedimentary material on the seafloor is deposited too slowly, and the blurring "bioturbation" effect of bottom-dwelling organisms is too great to capture the real pace of events revealed in Greenland.

The ice and the seafloor are sedimentary archives of very different processes. One is a record of turbulent atmospheric events laid down in annual laminations—snowfalls piled deep on the surface and slowly condensed into layers of nearly pure glacial ice. The other is dense, slurry muck—the very slow buildup of decayed matter and skeletal remains of marine organisms and fallout from melting icebergs. Like instruments performing different parts of a symphony, ice is playing the high staccato notes of the atmosphere while the seafloor sediment is grinding out the slow basal tones of the ocean.

Outside of Greenland, where in the world could climate archives that recorded changes on an annual scale be found? Researchers pursuing this question found themselves rummaging through the dustbin of geology, following a line of investigation that had been discarded by an earlier generation. They would embrace the very notion that earlier climate researchers had found so hard to accept—that far distant pieces of evidence could be related to one another. Oceanographers in particular were on the lookout for *varves*, the annually deposited "couplets" of sediment first identified and named by the Swedish geologist Gerhard de Geer early in the twentieth century. Stitching together distant events along a new timescale, paleoclimate researchers would rely on other concepts first given form by the Swede. It was de Geer who coined the terms *geochronology* to describe the pattern of time intervals illuminated by the layers and *teleconnection* to describe a linkage between far-flung events.

Two studies that first overcame the limitations of most ocean sediments were reported in 1992 by researchers from Norway and the United States. Their data, taken from the bed of the Norwegian Sea, north of the North Atlantic, produced marine climate profiles that were more closely akin to the pace and pattern of events seen in the Greenland ice. The cores were taken from an area of the sea in which biological activity deposited sediments much more rapidly than most regions of the world's oceans. An annual layering recorded the seasonal comings and goings of sea ice that first left the surface at the end of the ice age 13,400 years ago and thereafter opened an ocean corridor that was seasonally free of ice along the west coast of Norway up to 72° North.

Studying traces of algae, Nalân Koç Karpuz and Eystein Jansen of the University of Bergen analyzed two cores taken from the continental slope of Norway, a site that was critical to detecting important changes in North Atlantic circulation.

Scott J. Lehman and Lloyd D. Keigwin of Woods Hole Oceanographic Institution in Massachusetts studied a core taken from sediment just south of the Bergen team's core in the Norwegian Trench. Comparing their archive to the Dye 3 ice core, the Greenland standard at the time, the Woods Hole team said its "rates of deposition are comparable to those in the Greenland ice cores." Sediment accumulated in the Norwegian Trench at the rate of five meters every 1,000 years, 20 to 50 times higher than in previously studied North Atlantic cores "and much too high to be significantly affected by bioturbation." Lehman and Keigwin analyzed oxygen isotope variations and changes in the populations of cold-water species of plankton in the sediment to trace the movement and retreat of warmer North Atlantic water into the Norwegian Sea during the past 14,000 years.

The Norwegian Sea sediments revealed dramatic changes during the Younger Dryas and, for the first time, documented shifts in ocean currents that were as abrupt as the changes recorded in Greenland in the core taken at Dye 3. At the end of the ice age, about 13,400 years ago, the warmer Atlantic surface water began flowing into the Norwegian Sea. And 11,700 years ago, after about 1,800 years, the flow suddenly shut down again, just as the Greenland cores began picking up signals of an abrupt return to colder, drier, windier conditions.

In their study of the most rapid change recorded in ocean sediments, the Woods Hole team found evidence that "deglacial shifts in the poleward flow of warm Atlantic surface waters occurred extremely rapidly, typically within a 40-year period." Sea surface temperatures typically jumped 9°F. "These led to equally large and rapid changes in atmospheric temperatures," to shifts in Atlantic circulation and ice sheet melting rates. Directly downwind of these events, the alpine meadows of Scandinavia were intermittently blooming with the white flowers of *Dryas octopetala* as the Fennoscandian ice sheet waxed and waned.

The data added an important piece to the Younger Dryas puzzle, giving researchers a chain of events extending from the Greenland ice, through the ocean sediments of the far north, to the glacial moraines and terrestrial sediments of northern Europe. Other researchers would incorporate other processes into the scenario of sudden, high-latitude warming such as the shrinking of the Laurentide ice sheet, the meltwater flood into the North Atlantic, the subsequent collapse of the North Atlantic's circulation, and the sudden return of ice age cold to the region.

More generally, the Norwegian cores proved that abrupt change could be detected in certain ocean sediments. Together with the new ice core announcements in 1993, this "proof of concept" encouraged oceanographers to search beyond the continental margins of the North Atlantic for signs of the Younger Dryas cold reversal, the Dansgaard-Oeschger cycles, and the Heinrich events.

The obvious places to look for more annually deposited sediments were the submerged slopes of the continents, where seasonal rains or meltwater runoff or other regular changes over the nearby landscape leave telltale patterns on the seafloor. These continental margins are among the biologically most productive marine environments in the world, where the sea from top to bottom is alive with organisms. Researchers eventually would discover large areas of modern ocean shoreline where the seafloor intersects a biologically unproductive layer of ocean known as the "oxygen-minimum zone" that preserves layers in sediments. Varying in depth, the middle zone is a layer of water between the nutrient-rich depths and the highly productive surface region. Organic matter decays as it sinks, depriving the water column of oxygen necessary for the survival of bottom-dwelling organisms whose burrowing behavior obliterates the telltale layers of the sediment. Also especially valuable are areas in which the sediments are deprived of oxygen because the seafloor is isolated physically from the surrounding ocean by a sill, or submerged ridge.

Since the early 1990s, all of the world's oceans have yielded high-resolution sediments bearing the familiar signs of abrupt climate changes. As the signs of abrupt change became more and more global in scope, in fact, researchers began to doubt the original proposition that processes in the little North Atlantic were really responsible for all of this change. Among the most valuable sites are those in the North Pacific off the California coastline, in the northern Arabian Sea off the coast of Pakistan, in the tropical Atlantic north of Venezuela, and in the subtropics near the island of Bermuda.

In the eastern Pacific, the offshore Santa Barbara Basin already was well known for upper layers of sediment that recorded centuries of modern climate history. These layers had been used to study changes in fish populations and interesting short-term climate variations such as El Niño. At 600 meters below the surface, the

Santa Barbara sediments are deprived of oxygen by the presence of a surrounding sill that curbs ocean mixing in the deepest 100 meters or so. In 1992, the basin's potential as an archive of ancient climate was realized when the research vessel *JOIDES Resolution* of the Ocean Drilling Program extracted a 200-meter core dating back 160,000 years.

"It was a gold mine of scientific information on climate change," said marine geologist James P. Kennett. "Everybody was absolutely amazed at the quality of the site." In 1995, Kennett and B. Lynn Ingram reported that the annual layering from the runoff of winter rains was preserved in some sections of the core but obliterated by burrowing worms in other sections. They ascribed the interruptions to changes in ocean ventilation that altered the depth of the oxygen-minimum zone. As it happened, the pattern was just the signal that researchers were looking for, because it demonstrated the global reach of abrupt climate change.

In 1996, Kennett and Richard J. Behl compared the Santa Barbara Basin sediment pattern with the new GISP2 ice core record and found a close correspondence: The intervals when the laminations were preserved, when the basin bottom was anoxic, matched the record of Dansgaard-Oeschger warm events in the Greenland climate. It was what Kennett called "the first clear demonstration of Dansgaard/Oeschger cycles in another ocean, the biggest ocean, the Pacific Ocean, and at mid-latitudes."

A major turning point had been reached. "All of a sudden, everybody says, ah ha, this is a big process and it covers a large part of the globe," said Kennett. "And everything was so abrupt, it had to be through the atmosphere, the teleconnections. There was a fundamental change in the thinking about the importance of these D/O cycles after that in terms of the magnitude of these processes. Our work immediately showed that the surface oceans not only changed, but were capable of changing very, very abruptly, in the order of decades, just like Greenland. So the oceans were sort of in lock-step with the atmosphere as recorded in Greenland."

Later studies of sediments farther south, from the continental margin off the coast of Baja California, closely matched the Santa Barbara Basin results, although Joseph Ortiz and colleagues suggested that a different mechanism was at work. Kennett explained the changes in terms of changes in the ventilation of the seafloor by large-scale ocean currents. Ortiz pointed to more local changes in marine productivity between warm and cold periods, a process more likely to be the result of differences in wind patterns.

On the other side of the globe, different ocean and atmospheric conditions preserved annually layered tropical Indian Ocean sediments in the northeastern Arabian Sea that revealed a finely detailed profile of abrupt climate change. The continental shelf off Pakistan is a region of intense upwelling of nutrients and such robust biological productivity that the respiration of marine organisms periodically depletes the water of dissolved oxygen.

In 1998, a team of German researchers led by Hartmut Schulz presented sediment records from the Arabian Sea that are remarkably similar to Greenland climate oscillations over the past 110,000 years. When Greenland and North Atlantic temperatures were relatively high during the warm periods of a Dansgaard-Oeschger cycle, strong southwest monsoon activity led to high biological productivity; an oxygen-depleted Arabian seafloor; and dark, carbon-rich, well-preserved annual bands. Pale, carbon-poor disturbed laminations marked times of weaker monsoons and coincided with cold North Atlantic Heinrich events. Schulz wrote in *Nature* that these links between high-latitude and low-latitude climate events suggested "the importance of common forcing agents such as atmospheric moisture and other greenhouse gases."

Among the most extensive and valuable climate records to come out of an ocean floor core is the one from an area of the tropical Atlantic known as the Cariaco Basin in the southern Caribbean Sea off the coast of Venezuela. The Cariaco Basin is enclosed by especially shallow sills, at 146 meters and 120 meters, which isolate its deep water from ocean circulation. This means that the basin reflects only changes that occur in surface conditions, in the upper 100 meters of ocean. The changes are well preserved as laminations in the sediments because below 300 meters its waters are often depleted of oxygen. Deposits accumulate on the floor of the basin at the rate of 40 centimeters every 1,000 years, about 10 times faster than in the open Atlantic.

The north coast of South America is highly sensitive to climate change, and the basin sediments offer remarkably detailed records of features that are especially valuable in tracing the rapid changes revealed by the Greenland ice cores. The annual climate cycle, which leaves a distinctive "couplet" on the basin floor, is driven by changes in the strength of the northeast trade winds and the seasonal north-south migration of the rainy Intertropical Convergence Zone, where the merging northern and southern trade winds spawn great convective storms. A dry season from October to May is a time of strong trade winds and powerful coastal upwelling of organic nutrients from the deep ocean. The upwelling lays down a light-colored, organic-rich layer. A rainy season between June and October features weaker winds and no upwelling. Local river runoff leaves a dark band of mineral grains.

"The most striking large-scale feature of the Cariaco record is a significant increase in the thickness of the light laminae during the Younger Dryas cold period," oceanographer Alan E. S. Kemp told a meeting of the Royal Society in 2003. Kemp was describing the work of Konrad A. Hughen who compared their Younger Dryas data with results from the European ice core at Summit and found similarities even in small details. It seemed clear that high and low latitudes were responding to the same events, and Hughen's new radiocarbon studies confirmed that the Atlantic's overturning circulation of cold and warm currents was involved. The plunging return to cold conditions in the North Atlantic sharpened the temperature contrast between the high latitudes and the Tropics, increasing the strength of the trade winds, the upwelling along South America's northern shores, and consequently the thickness of the light-colored sediment layer composed of surface-dwelling marine organisms.

Other researchers extended the Cariaco Basin record back 90,000 years, closely measuring subtle changes in the reflectivity of the dark and light layers along 37 meters of sediment core. In 2000, Larry C. Peterson produced a climate profile through the last ice age that closely reproduces the abrupt comings and goings of the warm Dansgaard-Oeschger events recorded in the oxygen isotope profile of the U.S. GISP2 ice core.

Other researchers showed that the impact of Heinrich events extended down through the middle latitudes, well beyond the main belt of ice-rafted debris across the floor of the North Atlantic. Studying a varved sediment core from the continental slope of southern Spain, the French researcher Edouard Bard showed that "all latitudes of the eastern North Atlantic were strongly affected" by the three Heinrich events during the final 15,000 years of the last ice age. In the western Mediterranean, an analysis of varved sediments by Spanish sedimentologist Isabel Cacho found "a strong parallelism" between oscillations of sea surface temperatures and Dansgaard-Oeschger events, as well as five episodes of "drastic changes in the surface hydrography" that corresponded with Heinrich events. On the continents themselves, archives of abrupt climate change are found in varved sediments in the beds of lakes and layered mineral deposits in the stalactites and stalagmites in caves.

In southwestern France, inland of the Atlantic coast in a cave near Villars, Dominque Genty presented stable oxygen and carbon isotope profiles of a stalagmite that recorded D-O events during 50,000 years of the last ice age. In the eastern Mediterranean, studies by Miryam Bar-Mathews of stalagmites and stalactites in Soreq Cave in Israel, in the foothills of the Judean Mountains near Jerusalem, yielded a record of drought in the Middle East that corresponded with Greenland and North Atlantic records.

In China, a team led by Yongjin J. Wang reported in 2001 that oxygen isotope records of five stalagmites from Hulu Cave near Nanjing "bear a remarkable resemblance to oxygen isotope records from Greenland ice cores, suggesting that East Asian Monsoon intensity changed in concert with Greenland temperature between 11,000 and 75,000 years before the present."

By 2002, the global extent of abrupt climate changes was

documented by the German marine geologist Antje H. L. Voelker who compiled a database of 183 entries of climate records covering the period between 45,000 and 15,000 years ago. As the evidence accumulated, a more complex picture of climate changes emerged. The popular scenario of switches between warm and cold episodes driven by North Atlantic circulation changes seemed to satisfy most researchers as long as the rapid climate episodes were seen as regional in scope. Collapse or decline of the thermohaline circulation in the North Atlantic implied an opposite reaction in the south as world ocean currents reorganized the transport of tropical heat. What most researchers expected to observe was a "seesaw" pattern of cold-warm switches between the Northern and Southern Hemispheres. But this was not what they found.

Although the Southern Hemisphere is notoriously underrepresented in any compilation of world climate records owing to the predominance of ocean, the results clearly showed most climate changes moving in the same direction at the same time as Northern Hemisphere changes rather than opposite them. As researchers refined their ability to correlate the timing of changes among different records, this question of "synchrony" or "asynchrony" became central to understanding the mechanisms behind abrupt change.

In and around Antarctica, where ice generally accumulates more slowly and so leaves a less detailed climate profile than in Greenland, the evidence would prove especially confusing and contradictory. Ice core records at some locations were in opposite phase with the North Atlantic, supporting the seesaw idea, while records at other locations seemed to move in concert with the Northern Hemisphere. In 1998, European scientists led by the Swiss geophysicist Thomas Blunier synchronized the timing of large and rapid variations in the methane gas content of trapped air bubbles in both Antarctic and Greenland ice. Their comparison of the cores showed that during some large Dansgaard-Oeschger warming episodes, at least, Antarctic temperatures spiked upward more than 1,000 years sooner than temperatures in Greenland. On average, they reported, between 47,000 and 23,000 years ago, "Antarctic climate change leads that of Greenland" by 1,000 to 2,500 years.

Another notable study in 1995 by glacial geologist Thomas V. Lowell established the timing of the advances and retreats of glaciers in the Lake District of the Chilean Andes at 41° South and the Southern Alps of New Zealand at 44° South. Far distant from Greenland and large ice sheets, both adjacent to the Pacific and influenced by the midlatitude southern westerlies, they were what Lowell, writing in *Science*, called "prime localities for determining whether the North Atlantic climatic pulses were regional events or were part of a global signature." The evidence from both areas was clear: The glaciers in the midlatitudes of the south were moving in the same direction at the same time as the Northern Hemisphere ice sheets. As Lowell wrote, "The implication of global symmetry that arises from our Southern Hemisphere paleoclimate data underscores a fundamental lack of understanding of how rapid climate changes originated and were propagated globally."

Later, in 1999, a group of Swiss researchers led by Susan Ivy-Ochs and George H. Denton confirmed Lowell's conclusions with more precise dating techniques. Using the Swiss accelerator mass spectrometer, they measured the concentration of the cosmogenic isotope beryllium-10 at the exposed surfaces of boulders on the Lake Misery moraines at Arthur's Pass in the Southern Alps of New Zealand and an Egesen moraine complex at Julier Pass in the Swiss Alps. This study found "synchronous glacier advances both in the Southern Alps and in the European Alps during the European Younger Dryas."

In 2002, writing in the *Proceedings of the National Academy of Sciences,* Lowell noted that a similar pattern had been observed at widely scattered locations. "Such distribution implies an overall cooling of the atmosphere, not simply a regional redistribution of the heat balance," he wrote. From such a perspective, more investigations might answer more detailed questions about the timing of these cooling effects, and so "contribute to an explanation for the causes of abrupt climate changes."

By the end of the 1990s, the broad questions about the geographical spread of the major abrupt changes discovered in Greenland's ice had been resolved. With the possible exception of Antarctica, they seemed to be global in scope. At the same time, however, the search for a theory, for an explanation of the causes of abrupt climate change that was consistent with the data, did not prove as congenial to the methods of climate science.

10

NONLINEARITIES

How climate changes "faster than the cause"

The discovery that climate can change in just a few years can be seen as the culmination of a progressive ratcheting up of the pace of events as succeeding generations of researchers found better methods and tools that offered more accurate visions of the past. Certainly, change that might have been thought in the 1950s to take thousands of years was found in the 1990s to have occurred in a decade or less. But it is not just that things happen more quickly than expected. A climate that is subject to abrupt change is fundamentally different, more variable, and less predictable, posing questions that lead to different, more difficult explanations of causes and effects. Most researchers in the 1950s would have said that, aside from the far-distant ice ages, climate really doesn't change much at all.

The 1950s researchers were seeing climate from the perspective of a sailor observing a far shore. The contours of the distant land are smooth. In the grand scheme of things, on the planetary scale, the climate of Earth seems to have been relatively stable for the last 4 billion years at least. Various internal feedback processes keep temperatures within the narrow range that permits water to remain liquid—not so cold that it all freezes or so warm that it evaporates into space. The abrupt change pattern is the signature of the climate processes that keep this liquid window open. Seen for the first time on the human scale, the wiggles of climate change are much bigger, of course, more sudden, and not nearly so benign. Researchers in the 1990s were like sailors negotiating shallow coastal waters and making a landing. From the perspective of their launches pitching in the surf, the shore is lined with rocky crags and cliffs of abrupt change.

Recognizing the new detail and the new timescale of this more eventful history forces scientists across an uneasy theoretical divide. Left behind is the venerable idea that climate is a ponderous system anchored by stable, gradually evolving features—by the orbital motions of the planet, by the oceans and their long thermal memories, and by the ice sheets and their geologically paced advances and retreats. Such a system is complicated, to be sure, but not insurmountably so, and given time, its behavior is predictable. It is infinitely more congenial to analysis and simulation than a system driven by the dynamics of chaos. Abrupt change means that, like the weather itself, climate sometimes behaves in ways that defy prediction. Processes in the atmosphere, in the ocean, and on the land are known to interact with one another, and even though scientists think they know all of the parts and all of the important processes, still they cannot be sure of the outcome of these interactions from one time to the next.

Trying to capture these more elaborate and subtle but less predictable properties of their subject, climate scientists find themselves adopting a vocabulary that seems far removed from the mechanisms of climate and weather. It is the language used by computer modelers and mathematicians who study characteristics of chaotic and complex systems. Concepts with names such as *nonlinearities, feedbacks, critical thresholds,* and *multiple equilibria* are their stock in trade. Their formulations are designed to simulate the behavior of economic systems that provoke stock market crashes, biological systems that prompt mass extinctions, tectonic systems that cause earthquakes, and climate systems that change abruptly.

In the vernacular, the climate system is nonlinear, which means its output is not always proportional to its input—that, occasionally, unexpectedly, tiny changes in initial conditions provoke huge responses. It is chockablock with feedbacks, loops of self-perpetuating physical transactions, operating on their own timescales, that amplify or impede other processes. This constant cross talk of positive and negative feedbacks is said to be balanced, more or less, at various critical thresholds in the system. Forced across such a threshold, by whatever external or internal triggering mechanism, important variables begin gyrating or flickering, and the system suddenly lurches into a significantly different semistable mode of operation, a new equilibrium. All of these variables, all of these timescales, make for a system that is full of surprises. Scientists use an enigmatic phrase to describe how such chaotic systems violate one of the central principles of linear physics. *The whole is more than the sum of its parts.*

Nonlinear characteristics of the atmosphere and the ocean have been part of the literature since the early 1960s. In fact, it was a meteorologist who first described the physical principles of chaos. In 1961, in his office at the Massachusetts Institute of Technology, Edward N. Lorenz was using his little Royal McBee computer to execute a rudimentary numerical model of the atmosphere. The slightest changes in initial conditions, Lorenz discovered, inevitably led to major changes in the output of the model as the computer performed repetitions of his equations. As Lorenz and others developed the theory, they laid the groundwork for the way scientists in a variety of fields study the behavior of dynamic systems.

At the same time, just down the way at Woods Hole Oceanographic Institution, oceanographer Henry Stommel was developing the idea that the North Atlantic had two different patterns of circulation and that it was capable of abrupt transitions between these equilibrium states. At the time, this concept seemed to conform to the prevailing theories about climate changing between ice ages and warm periods in concert with long-term changes in the geometry of Earth's orbit of the Sun. It was Stommel who also introduced the concept of *thermohaline circulation* that linked the motion of currents to differences in water density, a process driven by changes in temperature and salinity. It would be many years, not until the mid-1990s, before climate scientists knew what to make of these ideas.

An abrupt climate change is the very definition of a nonlinear event. In 2002, a report by a special committee of the National Research Council (NRC), chaired by Richard Alley, so defined it: "Technically, an abrupt climate change occurs when the climate system is forced to cross some threshold, triggering a transition to a new state at a rate determined by the climate system itself and faster than the cause." Moreover, the cause of such an abrupt climate change may be "undetectably small."

Change faster than cause and cause undetectably small—these are the footprints that Lorenz followed through the curious behavior of his computer simulation. This discovery had fairly obvious and discouraging implications for weather forecasters, who in the 1950s had dreamed of the day when the usefulness and accuracy of their predictions would extend far into the future. Now they know that, although ingenious numerical models executed by the most powerful computers generate excellent simulations of typical atmospheric changes, capturing the nonlinear motions of thermodynamics, still there are practical limits to their weather forecasting skill. Because the system is chaotic, a typical three-day forecast is not nearly as reliable for the third day as for the second.

When they come to simulating climate, powerful computer models seem to encounter the same sorts of problems with chaotic behavior in the system. Throw all of the interactions and feedbacks into a model that couples the intricate physics of the atmosphere and the ocean, and the output fairly accurately reproduces the typical workings of the system. But that is not always the way the world works. "There have not yet been any successful simulations of the pattern and magnitude of the Younger Dryas or of recurrent Dansgaard/Oeschger events with coupled ocean-atmosphere general circulation models," Alley and colleagues wrote in the NRC study. Nobody is sure why this is so, but something important seems to be missing. Theorists point to problems with models, and modelers point out that the theory could be wrong.

Summarizing their work in a 2003 review in *Science*, the NRC team described a consistent "mismatch" between the computers' idea of abrupt climate change and the real record of the last 100,000 years. Across the board, the real record shows changes that were greater and more widespread than the computers reproduce.

Is it possible that the computer models have it right and climate scientists have it wrong, that they are misinterpreting their data? Not likely, say the scientists. The record of the past is too clear, the cross-checks are too numerous to take such doubts very seriously anymore. That argument collapsed when the big Greenland projects reported their polar ice findings from Summit. Either some natural processes are missing from the models, the NRC team concluded, or the computers systematically "underestimate the size and extent of climate response to threshold crossings."

The difficulties of simulating past abrupt climate changes lead some researchers to question whether predicting future climate is even a realistic goal. Participants in a 2001 workshop at Duke University on nonlinearity in the environment posed the question: Given the nearly certain occurrence of sudden transitions between climate states, is "prediction" per se achievable? The group seemed impressed most by "a relatively poor understanding" among researchers of the nonlinear character of the climate system and the mechanisms that drive it to rapid, episodic change. "Abrupt climate change is believed to be the result of instabilities, threshold crossings and other types of nonlinear behavior of the global climate system, but neither the physical mechanisms involved nor the nature of the nonlinearities themselves are well understood," wrote theorist José A. Rial, of the University of North Carolina's Chapel Hill Wave Propagation Laboratory, and colleagues in the journal *Climatic Change* in 2004. Citing several examples of nonlinearities, the group was led "to an inevitable conclusion: since the climate system was complex, occasionally chaotic, dominated by abrupt changes and driven by competing feedbacks with largely unknown thresholds, climate prediction is difficult, if not impracticable."

Whether prediction will ever be accurate or not, it looks, for better or worse, as though computer modeling and the study of abrupt climate change are wedded to one another. It is a weird marriage of extremes. In the antiseptic, environmentally controlled surroundings of computer laboratories, mathematicians manipulate ice core and ocean sediment data that have been gathered in some of the coldest, wettest, most difficult heavy-lifting fieldwork in science. And yet, in the study of climate, it is hard to imagine one line of research without the other. Weather forecasters don't have long to wait to find out if they've got the system right, but without computer simulations, climate researchers really have no way to test their ideas. More than that, however, although the evidence for abrupt change now seems obvious, there is nothing obvious about the explanations for it. The old idea of a stable, slowly evolving climate was so widespread through the twentieth century—and died such a slow death—because it seemed to make the most sense. Abrupt change, in contrast, is so counterintuitive and so elusive that it is like a concert being played at a pitch beyond the range of human hearing. The chaotic nature of the climate system might never have been recognized were it not for the mathematical genius of Edward Lorenz and his little Royal McBee computer. Now as then, it is the so-called simple computer models or models of "intermediate complexity" that are most useful for testing the validity of one idea or another.

Another instance of the merging of meticulous data analysis and computer modeling has led to interesting evidence that the curious nonlinear behavior known as "stochastic resonance" may account for a 1,500-year cycle of abrupt change detected in the record. It may explain the timing of the two dozen Dansgaard-Oeschger events that dominate the last 100,000 years of climate.

The North Atlantic's nonlinear thermohaline circulation—its density-dependent formation of deep water in the far north, its abrupt switching between multiple modes of equilibrium—remains a central feature of the leading rapid change scenarios, although not all of the prominent events during the last ice age are likely to yield to the same explanation. After all, the most widely observed and carefully studied rapid change—the Younger Dryas, 11,400 years ago—was a sudden plunge back toward frigid cold conditions that came at a time when Earth was warming out of the last ice age. In contrast, the most common episodes of abrupt change in the past 110,000 years, the Dansgaard-Oeschger events, were periods of sudden warming that came with some regularity throughout the ice age.

To explain the Younger Dryas, theorists can point to specific places, times, and events. About 14,000 years ago, as Earth began warming from the last ice age and the great Laurentide ice sheet began shrinking, a vast meltwater lake began to form at the edge of the ice in southeastern Canada. The Great Lakes are remnants of this ice sheet melting. Sediments in the Gulf of Mexico attest to this meltwater lake draining for centuries into the Mississippi River. Then, about 11,400 years ago, so the story goes, an ice dam gave way and an enormous flood of fresh water suddenly poured through the St. Lawrence River into the North Atlantic. This rapid freshening so altered the density balance of the ocean that far northern surface water no longer sank to the bottom when it chilled. This halt in deep-water formation caused a shutdown of the ocean thermohaline conveyor, closing off the flow of warm water from the Tropics into the North Atlantic and plunging the region back toward ice age conditions. A cold, dry, windy regime held sway for 1,300 years before warming resumed. The Younger Dryas cold was transported to other regions by circulation changes in other oceans and broadcast by large-scale and persistent atmospheric responses, paleoclimatologists believe, leaving a distinct climate signal over much of the globe.

But what of the main features of ice age climate history? What caused the Dansgaard-Oeschger warmings and the supercold Heinrich episodes? A group of leading abrupt-change specialists pondered these questions at an American Geophysical Union conference in 1998 in Snowbird, Utah. The stabs of extreme cold—the six Heinrich events during this glacial period—seem closely tied to the D-O cycle, coming as an exaggerated cooling phase at the end of a bundle of several warming events, each separated by periods of progressively colder temperatures. Evidence from the ocean sediments, showing the Canadian origin of the ice-rafted debris layer, pointed to the dynamics of the Laurentide ice sheet as their cause. Most researchers have accepted the idea proposed by glaciologist Douglas R. MacAyeal that the ice sheet over North America underwent a binge-purge cycle, becoming unstable as it built up over thousands of years. MacAyeal proposed that Heinrich events "were caused by free oscillations in the flow of the Laurentide ice sheet which arose because the floor of Hudson Bay and Hudson Strait is covered with soft, unconsolidated sediment that forms a slippery lubricant when thawed." The great ice sheet "periodically disgorged icebergs in brief but violent episodes which occurred approximately every 7,000 years," MacAyeal said, when geothermal heat trapped under the overburdened glacier thawed its base. These icebergs pushed the spreading sea ice to exceptionally low latitudes, amplifying the cold conditions, completely shutting down the conveyor, and halting formation of deep water in the North Atlantic.

As Alley summarized the consensus view in an AGU monograph, the Dansgaard-Oeschger events, while clearly related to ocean circulation, were a bigger puzzle. The big climate swings have come not when conditions such as temperatures and ice extent were high, and forces such as carbon dioxide in the atmosphere or radiation from the Sun were extreme, but rather when conditions were relatively moderate and CO_2 and solar forces were changing rapidly. Maybe the swings came when forces would push the climate system "in the gap between two modes." Alley offered an alternative explanation—that "the stability of the system may be sensitive to the rate of change of forcing variables, such that climate response is analogous to that of a drunken human: when left alone, it sits; when forced to move, it staggers with abrupt changes in direction."

What could provoke the sudden warmings in the midst of an ice age? In his lectures around the country, Alley found that his ideas attracted other scientists to his line of thought. The Penn State glaciologist soon found himself teamed with two new partners, geophysicist Sidhar Anadakrishnan and physicist Peter Jung, to test a line of thought that stochastic resonance might account for what seemed to be a 1,500-year period between most D-O events. The idea behind stochastic resonance is that a feeble periodic signal is amplified by the presence of stochastic—that is, random—"noise" in the climate system, that the weak signal and the noise resonate and produce a signal capable of triggering an abrupt change.

Alley performed a frequency analysis of the oxygen isotope signals of the U.S. GISP2 and European GRIP ice cores. If stochastic resonance played a role in the 1,500-year cycle, the researchers reasoned, a certain signature would emerge from frequency analysis of the climate record. If the cycle were strictly periodic—strictly linear—all transitions would land on the 1,500-year mark on a histogram, a chart depicting frequency distributions. If it were no more than random noise at work, events would be scattered over the histogram. What emerged on the histogram was just the pattern that stochastic resonance would leave. Almost all events landed at multiples of 1,500 years—most at the 1,500-year frequency, fewer at 3,000, and fewer still at 4,500. While other researchers confirmed that the North Atlantic's conveyor system was potentially subject to stochastic resonance, they estimated that only a relatively powerful signal would be able to switch it on and off.

While Alley and coworkers were massaging their data, Stefan Rahmstorf and Andrey Ganopolski were doing some interesting computer modeling at the Potsdam Institute for Climate Impact Research in Germany. Their "intermediate complexity" model simulates a more sensitive ice age North Atlantic than other models of the ocean. The warm mode of their model Atlantic, when warm water is flowing northward, cooling, and sinking to the depths, is an "excitable" state that is less stable than the glacial mode. As the German researchers put the case in a 2001 monograph, "we have shown that the 'cold' mode is the only stable mode of the glacial Atlantic ocean circulation." Rather than completely switching off deep-water formation, their model shifts the critical overturning zone farther south. Alley's team noticed that this ice age ocean is more likely to respond to stochastic resonance.

Alley and Rahmstorf brought their two lines of research together in an article in the American Geophysical Union journal *Eos.* Combining random noise with a very weak 1,500-year climate cycle in this model produces Dansgaard-Oeschger events that "are very similar to those recorded in Greenland ice and other paleo-climatic archives," they wrote. And the recurrence times closely resembled the pattern developed by Alley and colleagues. "This shows how the Atlantic ocean currents can act like a threshold amplifier, turning a feeble signal into dramatic climate swings."

More work was going to have to be done before anyone could confidently claim that stochastic resonance explains abrupt climate change, they wrote. "But for the first time, data analysis has shown strong hints and model simulations have provided a possible quantitative mechanism for stochastic resonance causing some of the most abrupt climate shifts known." The concept raised interesting questions. What is the source of the 1,500-year cycle? Could natural or human-caused noise affect the system in the future?

Several researchers have proposed possible causes of a weak periodic signal. As early as 1990, Wally Broecker had suggested that abrupt climate events were triggered by oscillations in the salt content of the Atlantic that periodically shifted the density balance of the ocean and switched the circulation conveyor on and off. In 1997, the U.S. ice core project leader at Summit, Paul A. Mayewski, analyzing the chemical properties of the GISP2 ice core, described a 1,450-year cycle of rapid climate change events that involved "massive reorganizations of atmospheric circulation." The glaciologist suggested that solar variation ice sheet-atmosphere feedbacks may help explain the timing and amplitude of rapid climate changes. That same year, Gerard Bond at Lamont-Doherty, analyzing North Atlantic sediment cores, found a long cycle of abrupt change on a timescale of 1,470 years that ran through the ice age and—at lower magnitude—continued to punctuate the supposedly stable climate of the Holocene, the last 11,000 years. In 2001, Bond and colleagues also proposed that these cycles, marked by the spread of drift ice in the North Atlantic, were tied to solar variation. In 2000, Charles D. Keeling and Timothy P. Whorf suggested that a 1,800-year tidal cycle was responsible for abrupt climate change. According to them, "variations in the strength of oceanic tides cause periodic cooling of surface ocean water by modulating the intensity of vertical mixing that brings to the surface colder water from below."

Apart from the enormous fact of it—the data in the ice cores and the ocean sediments that plainly show climate changing dramatically in just a few years and the increasing evidence of its global reach—almost no important issues about the mechanisms of abrupt climate change are entirely settled. The problem is new; the science is young. Still scientists are asking themselves questions as basic as that posed in 2003 in the journal *Science* by Wally Broecker: "Does the Trigger for Abrupt Climate Change Reside in the Ocean or in the Atmosphere?" Focusing on the Younger Dryas, the most thoroughly documented of the abrupt changes, Broecker reviewed data suggesting that the North Atlantic's circulation was responsible and other data favoring an atmospheric mechanism centered in the equatorial Pacific. The oceanographer found in favor of the ocean, again, and the scenario of change driven by reorganization of the ocean conveyor in the North Atlantic that he first proposed in the mid-1980s. Still, the state of the science is such that few conclusions can yet be stated with much conviction, even by a theorist who is famous for doing just that. As Broecker would observe, "we are still a long way from understanding how our climate system accomplished the large and abrupt changes so richly recorded in ice and sediment."

The idea that the oceans play a central role in the climate of a planet so dominated by the presence of water is not really in dispute. At every timescale, the atmosphere and the ocean are engaged in intimate conversation—the atmosphere picking up water vapor and temperature changes and discharging precipitation and current-bending winds, the oceans storing and transporting heat from the Tropics and cold from the poles. Coming from Broecker, the question of a trigger or mechanism for change in the ocean or the atmosphere is recognition of an ongoing controversy among leading theorists and modelers about where in the world abrupt climate change originates. The scenario that invokes the thermohaline circulation and its mode shifts in the North Atlantic was the first and still is the most thoroughly developed line of thought about the causes and the timing of abrupt changes through the past 100,000 years.

Broecker would be the first to acknowledge that the visual icon of this reasoning, the Great Ocean Conveyor of continuous strips of cold and warm currents wending their ways through the world's oceans, is a cartoon rendering of how the oceans really work. First drafted for a popular magazine article in 1987, the picture would forever offend physical oceanographers for its oversimplification. "Had I known that this conveyor belt diagram would become a logo for global change research," he would recall, "I probably would have worried a bit more about the details."

More fundamental are differences in evaluating the role of thermohaline circulation—the THC—the currents that respond to density differences that are themselves a function of temperature and saline properties of water at different depths. In a 2000 commentary in the *Proceedings of the National Academy of Sciences,* the German theorist Jochem Marotzke wrote: "The THC's role in abrupt climate change is not comfortably established—on the contrary, it poses major scientific challenges and thus provides a powerful focus for climate research." Elsewhere, Marotzke observed that "first-order advances are needed" to fill critical gaps in the basic concepts of how climate changes. "Our current knowledge is insufficient to determine the likelihood of abrupt climate change, or even to answer some simple questions with confidence," he wrote. "For example, why was the THC in glacial times apparently subject to large instabilities, but in the present interglacial has been considerably more stable? Did the giant ice sheets dramatically amplify a climate signal that has persisted into the interglacial, but at a suppressed amplitude? Or does the relative quiescence of the present interglacial primarily result from a change in the atmospheric forcing applied to the ocean, or perhaps a change in ocean processes, the nature of which may have altered because of decreased sea ice coverage or a higher sea level?"

Oceanographer Carl Wunsch of the Massachusetts Institute of Technology argues that the THC concept has been overused and overstated. In a 2002 article in *Science*, Wunsch maintained that the ocean's motions are "sustained primarily by the wind, and secondarily by tidal forcing." In both models and the real ocean, the buoyancy of surface water strongly influences the flows of heat and salt, "because the fluid must become dense enough to sink, but these boundary conditions do not actually drive the circulation."

The following year in *Nature*, modeler Stefan Rahmstorf noted that the THC "can be defined as currents driven by fluxes of heat and freshwater across the sea surface and subsequent interior mixing of heat and salt." This process is "clearly distinct" from the turbulent mixing caused by winds and tides, although because they work on the same circulation, the two mechanisms cannot be separated from one another in ocean measurements. In any case, heat-transporting ocean currents are critically important to the climate of regions surrounding the North Atlantic. "Changes in these currents are our best explanation for the abrupt and marked climate swings that occurred over the north Atlantic many times during the most recent glacial period, as shown by Greenland's ice cores and by deep-sea sediments," Rahmstorf wrote.

Changes in North Atlantic ocean circulation may be the best explanation, but there are other explanations for abrupt change—for the triggers, amplifying feedbacks, and "globalizer" processes that transmit its impacts across the planet. The gaps in knowledge still are large enough to accommodate more than one school of thought. Theorists and modelers are like frustrated cooks trying to work up a delicate soufflé without the benefit of a satisfactory recipe. Try after try, they can't get the dish to come out they way they want. Perhaps the problem is more than a matter of tweaking the proportions of this or that ingredient. Maybe it's not a soufflé recipe at all.

What bothers many scientists about the North Atlantic THC scenario of abrupt change is not so much what it proposes—that changes originate in the North Atlantic—but what it proposes to do without—the tropical Pacific. In the scheme of things, one might think it would be difficult to account for such powerful environmental episodes without involving the world's largest ocean or the Tropics, where sunlight deposits most of the planet's energy. Nevertheless, theorists were much influenced by 1970s research that seemed to show that while the rest of the world slipped into the last ice age, the Tropics cooled only a mere 1.8°F. Is the formation of North Atlantic deep water really the key to abrupt climate change? Or is it just the explanation most readily at hand that climate scientists happened to find?

Being a human enterprise, science is not always such an orderly business. Would our current understanding of natural climate variation be the same if Henri Bader had not persuaded the U.S. National Committee for the International Geophysical Year to support an expedition to Greenland in the 1950s? What if stable isotope technology had not been available in the 1960s to tease the temperature of ancient snowfall out of the polar ice? Climate science seems especially dependent on technology, its direction subject to the whims of contingency. The Swiss climate modeler Thomas F. Stocker touched on this issue in *Science* in 1998 as he evaluated prevailing thinking about the causes of abrupt change. "One of the pressing questions is where the centers of activity responsible for these abrupt and millennial-scale climate changes are located," he observed. "The region of the high northern latitudes, especially the North Atlantic, has been the classic focus of this research. However, it may be argued that the first institutions of paleoclimatic research have been located around the North Atlantic, and a certain bias cannot be excluded."

The same sort of bias probably could be observed in most lines of research. Lecturing to undergraduates, Harvard University geneticist Richard Lewinton tells the old story of the drunkard whom a passerby finds crawling around on his hands and knees under the light of a street lamp. He's looking for his car keys, he says.

"So, you lost your keys down here somewhere?" asks a passerby.

"No, I dropped them across the street," says the drunkard, "but this is where the light is good."

Maybe it is just a matter of proportion. Later research found much more intense cooling of the Tropics during the last ice age—locally as much as 7.2°F. And while no one can yet describe the details, no one really argues that the Tropics have nothing to do with abrupt climate change. In 1997, for instance, even as he was describing thermohaline circulation as "the Achilles heel of the climate system," Wally Broecker was ruminating about the climate-altering power of water vapor, the major greenhouse gas that is primarily a product of the Tropics. As Alley put the case in a commentary in *Nature* in 1998, "The idea that changes at high latitudes can affect widespread regions extending across the Equator is unpopular with many workers who do not believe that the small, energy-starved polar 'tail' can wag the large, energy-rich tropical 'dog.'"

In 1999, speaking for the tropical dog, geophysicist Raymond T. Pierrehumbert wrote in the *Proceedings of the National Academy of Sciences*: "It is not certain that the millennial scale fluctuations observed in different parts of the globe are all related to each other, but it is certain that such fluctuations are observed just about everywhere, so one should at least entertain the possibility that the NADW [North Atlantic deep water] picture alone cannot account for everything that is going on." It doesn't explain climate fluctuations in different regions, for instance, or how the impact of changes in North Atlantic deep-water turnover is spread around the globe.

Pierrehumbert was a member of the National Research Council's Committee on Abrupt Climate Change whose 2002 report described the tropical Pacific, where the waxing and waning of the El Niño-Southern Oscillation spreads well-known patterns of deluge and drought around the world on a three- to five-year cycle, as "a natural candidate to be a 'globalizer' of climate influences." Maybe it is more than that. The atmosphere's response to the tropical Pacific's main features—its wind-driven eastern "cold tongue" and western "warm pool" and the line of intense convection known as the inter-tropical convergence zone—brings to Pierrehumbert's mind the same sort of dynamical behavior that Broecker envisions in the ocean currents of the North Atlantic.

"The formation of convecting air is tippy in precisely the same sense and for precisely the same reason that formation of North Atlantic deep water is tippy," Pierrehumbert wrote. The tropical atmosphere is a powerhouse of storminess, but the air itself—like North Atlantic surface water—is delicately balanced, rising in response to a subtle interplay of moisture and temperature changes. As Pierrehumbert put it, "Tropical convection creates 'top air' in much the same way oceanic convection creates 'bottom water.'"

"The tropical atmosphere-ocean system offers a rich palette of possible amplifiers and switches that could in principle lead to abrupt climate change," the NRC committee wrote. Foremost among them is water vapor, the greatest greenhouse gas, what Pierrehumbert called "the premier atmospheric feedback amplifying the response to what would otherwise be minor climate forcings." Moreover, the circulation of the Tropics produces "boundary layer clouds" that can have an albedo—that is, reflectivity—"approaching that of ice and exert a potent cooling effect when they form. . . . As a nexus for climate change, the tropical Pacific also has a clear advantage over the North Atlantic, in that the global atmosphere is exquisitely sensitive to tropical Pacific sea surface temperature anomalies," wrote Pierrehumbert.

Variations in the surface salinity of the tropical and subtropical Atlantic during the past 136,000 years have been linked to the waxing and waning of the North Atlantic's conveyor by University of California researchers who analyzed two sediment cores from the Caribbean Sea. Cold periods bring relatively dry conditions to the tropical Atlantic, so the waters of the Caribbean become increasingly salty. How much the North Atlantic's circulation picks up during the next warm period may depend on how much extra saltiness the Caribbean has to deliver. Reporting these results in *Nature* in 2004, Matthew W. Schmidt and Howard J. Spero surmised that the tropical rainfall cycle "has a direct role in regulating rapid climate change."

Could El Niño trigger abrupt change? In modern climate, at least, nothing more dramatically rearranges the hydrologic cycle of the Tropics from one year to the next. In Germany, Mojib Latif tested the effects of warming ocean temperatures and was able to show a link between the salinity of the tropical Atlantic and El Niño-like changes in air-sea interactions in the tropical Pacific. Moving the center of heavy precipitation into the central tropical Pacific creates a warm, dry regime of descending air over northeast Brazil that increases evaporation and salinity in the surface of the tropical Atlantic. Larry C. Peterson studied sedimentary records from the Cariaco Basin off northern Venezuela and found evidence that "supports the notion that tropical feedbacks played an important role in modulating global climate" during the last ice age. In 2003, David W. Lea, Peterson, and colleagues compared the timing of sea surface temperatures recorded in Cariaco Basin sediments and suggested an "active tropical feedback" between the western tropical Atlantic and Greenland air temperatures during the Younger Dryas.

Modeling work by Amy C. Clement and coworkers points to the El Niño-Southern Oscillation as a possible trigger for abrupt climate change. In response to subtle, constant alterations in solar radiation, this unstable tropical Pacific ocean-atmosphere cycle can lock "in phase" with the seasonal cycle, producing several centuries of La Niña conditions. Their experiments suggest that such a locked-in La Niña could have triggered the Younger Dryas about 11,000 years ago.

James P. Kennett, investigating the Santa Barbara Basin, has developed an entirely different line of thought about abrupt change mechanisms. He challenges the common view that variations in the concentration of methane gas in ice cores and other climate archives are primarily signals of the waxing and waning of low-latitude wetlands as temperatures and sea levels rose and fell. This provocative "Clathrate Gun Hypothesis" argues that the powerful greenhouse gas itself was the trigger mechanism for abrupt warmings. Crystalline gas hydrates embedded in sediments along the continental shelves, sensitive to changes in temperature and pressure, occasionally became destabilized and produced bubbles that rose to the surface and released massive amounts of methane gas into the atmosphere.

Methane-laced ocean sediments represent an enormous reservoir of carbon—by some estimates, at least twice the amount of carbon found in all known fossil fuels. Warren T. Wood has shown that hydrates in some continental slopes may be more unstable than previously thought. In 2003, Kai-Uwe Hinrichs, examining the Santa Barbara Basin sediments, studied fossil remnants of bacteria that would have flourished only under high concentrations of methane. "It is evident from our data that substantial quantities of methane were rapidly mobilized at least several times over the past 60,000 years," Hinrichs reported in *Science.* Hinrichs's research supports the Clathrate Gun Hypothesis, although it doesn't say how much methane escaped from the ocean. "But one thing is for sure," he said, "our results clearly show that relatively minor environmental changes can have a major impact on sensitive coastal regions with yet unknown consequences for climate and [for life on Earth]."

11

HOLOCENE

The climate crash that rocked the cradle of civilization

Some 4,300 years ago, circa 2300 BC, an era that archaeologists call the Bronze Age, any rich and thriving civilization must have looked to all the Old World like a fixture of permanence and security. The transformation from subsistence farming to organized irrigation agriculture had given rise to culture and organized economies across a large swath of Earth from the Levant to Asia. Cities were large. Trading networks were prosperous. Classes of professionals were engaged in writing and mathematics, and nobles were articulating public policy.

In northern Africa, along the Egyptian Nile, the pharaohs of the Old Kingdom had built the pyramids at Giza. In Mesopotamia, between the Tigris and the Euphrates, humanity's first empire, that of Sargon the Akkadian, sprawled over 800 miles from the Persian Gulf to the northern Mediterranean. In southern Asia, across an area larger than Europe, the Harappan civilization of the Indus Valley had invented writing and employed artisans and traders in large and prosperous cities.

And then they were gone. The Harappan of the Indus Valley abandoned their large cities of Mohenjo-Daro and Harappa while northern India grew in population. In the "post-urban" farming culture that survived, writing lay fallow for 1,400 years. So steep was the decline of the Old Kingdom that according to an inscription on the tomb of Ankhtifi, a regional governor in southern Egypt, "all of Upper Egypt was dying of hunger to such a degree that everyone had come to eating their children." The Akkadian empire was no more.

What in the world happened? Pondering this question for more than a century, generations of archaeologists and Near East scholars have been embroidering on orthodox historical and anthropological explanations about economic, political, and military upheavals of one kind or another. Societies are complex systems, after all, subject to any number of internal and external forces and to their own nonlinear behavior. For one reason or another, the idea that climate changes played an important role in this history has not been a very big part of the mix of answers. Since the 1990s, however, advances in paleoclimatology and other earth sciences have been forcing themselves on this question. And the evidence continues to accumulate: Abrupt climate change took hold of humanity's first great civilizations and shook them until they collapsed.

Only recently have researchers come to realize that the last 10,000 years—the epoch known as the Holocene—which saw the rise of humanity, has not been the benevolently stable climate that scholars and scientists have assumed. And they are beginning to tap the potential of Greenland ice and other archives to shed new light on events on the scale of human history. On this scale, of course, it is not the global but the regional climate changes that push humanity around. Hidden inside the variable of global temperatures is the more powerful circumstance of their *differences* between one place and another. Climate scientists know that changes in these temperature differences alter the circulation of the atmosphere, and this is where climate hits the pavement of human experience. This is what is most important to societies: not the temperature changes themselves, but how these changes affect precipitation patterns over time—where in the world it rains or snows and how little or how much. In the 1990s, research by ecologist Kathleen R. Laird of sediments in Moon Lake, North Dakota, developed a striking record of abrupt, large, and long-lived changes in wet and dry conditions over North America during the last 2,000 years. Societies large and small are resilient and almost—but not quite—infinitely resourceful. Any society—however large or small, however resourceful—can reach a point, a threshold at which it is exquisitely sensitive to the availability of water.

In the 1980s, archaeologist Harvey Weiss was studying a site called Tell Leilan in northern Mesopotamia when he hit on the connection between abrupt climate change and the collapse of Old World civilizations. Tell Leilan is in a part of the Near East called the Habur Plains, where humans have been cultivating barley and wheat since the Neolithic Natufians first took up farming 8,000 years ago. As Sargon the Akkadian extended his reach in 2300 BC, these grain fields of the northern Habur Plains would become the breadbasket for his empire.

During the 1980s, Weiss was developing a theory around a pattern he was noticing in the excavations. The abandoned cities and settlements of northern Mesopotamia were in a region that, then as now, enjoyed a Mediterranean climate of wet winters and dry summers. Farmers depended on rainfall to water their crops. Yet, suddenly, for some reason, by the thousands, Akkadian citizens migrated downstream to southern Mesopotamia, where irrigated agriculture was employed.

In 1993, Weiss and colleagues reported finding in the dust of abandonment in Tell Leilan, site of the ancient Akkadian city of Shekhna, evidence for the sudden onset of drought that struck in 2200 BC and lasted 300 years. After years of conjecture among historians and archaeologists about the role of climate in the rise and fall of civilizations, Weiss's research had reached a turning point, a potent new collaboration. Coming just as the Greenland ice core projects were confirming the reality of abrupt climate change and hinting at its global extent, the Weiss studies pioneered a fascinating reevaluation of two very different lines of research. Archaeologists began seeing events in terms of climate, and paleo-

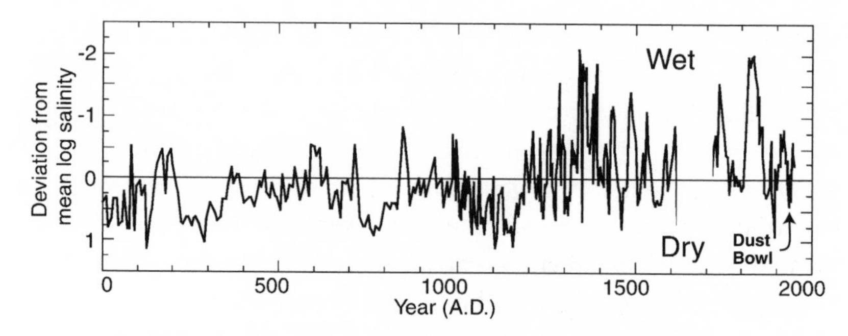

Drought Record from Moon Lake, North Dakota

A reconstruction from studies in the 1990s of fossil algae in sediments from this High Plains lake by Kathleen R. Laird and colleagues shows many long, intense droughts during the last 2,000 years. Notice the abrupt change in the pattern about AD 1200. The famous "Dust Bowl" drought of the 1930s appears to have been relatively minor and brief compared to other episodes in this record. Other climate reconstructions, such as tree ring records in California, show that these droughts were widespread.

climatologists began seeing episodes of Holocene change in terms of human history. At the Shekhna excavation was the French geologist and soil scientist Marie-Agnès Courty, who had pioneered new techniques that proved especially valuable at Tell Leilan. Courty provided the hard data that made the case.

Among the layers of earth representing successive eras of habitation at the site was a large section nearly two feet thick that dated from the time of the abandonment of Shekhna in 2200 BC. Unlike the rest of the layers, this sandy material contained no organic matter and showed no wormholes or other indications of moisture.

Powerful confirmation soon came in the work of paleoceanographers Heidi M. Cullen and Peter B. deMenocal, who analyzed the dust in a sediment core taken from the Gulf of Oman some 1,300 miles downwind of Tell Leilan. Isotopic analyses identified the dust as Mesopotamian. The chemistry of a volcanic ash layer in the ocean sediment matched the tephra layer recovered at Tell Leilan. And, according to Cullen and deMenocal, the calibrated dates of "the aridification and social collapse events are indistinguishable." In 2000, in an article in the journal *Geology*, the team concluded: "All available evidence indicates that this event records a dramatic mid-Holocene change in regional climate. . . ."

The team pointed to a variety of other paleoclimate records from the Middle East that document a sudden shift to dry, windy conditions across the region at the same time. In Lake Van in eastern Turkey, at the headquarters of the Tigris and Euphrates Rivers, sediments recorded a 30- to 60-meter fall in the lake level. To the south, the level of the Dead Sea fell by 100 meters. Lake levels fell in East Africa and North Africa as far west as Morocco.

Everything implicated abrupt climate change as "a key factor" in the collapse of the Akkadian empire. As Weiss had surmised, immigrants fleeing the famine to southern Mesopotamia swelled the cities beyond their capacity. Deprived of grain from the north, the Akkadians found their resources overwhelmed. They built a 100-mile wall against the "invaders" from the north, but soon the highly sophisticated society collapsed in violence and anarchy. From beginning to end, the imperial empire had lasted only a century. "Furthermore," wrote Cullen and deMenocal, "these responses occurred despite the fact that the Akkadians had implemented sophisticated grain-storage and water-regulation technologies to buffer themselves against historical . . . variations in rainfall."

Using the same northeastern Arabian Sea sediment cores, geochemist Michael Staubwasser examined the link between the abrupt change 4,200 years ago and the collapse of the Harappan civilization in the Indus River Valley. Oxygen isotope shifts in a sediment core revealed a sharp decline in the outflow from the Indus River 4,200 years ago that transformed the Indus Valley civilization from "a highly urban phase to a rural post urban phase. In particular," wrote Staubwasser, "cultural centers, such as the large cities of Mohenjo-Daro and Harappa, were almost completely abandoned while locations in northern India grew in population." The coincidence of the Harappan migrations and the drop in Indus River discharge 4,200 years ago was hard to ignore. "A possible explanation is that a reduction of the average annual rainfall over the Indus river watershed restricted Harappan farming in the Indus valley and left large city populations unsustainable."

Along the Nile, meanwhile, crop failure and famine that struck the Old Kingdom ended the 90-year reign of Pepy II in 2152 BC. "Within the span of 20 years, fragmentary records indicate that no less than 18 kings and possibly one queen ascended the throne with nominal control over the country," the historian Fekri Hassan wrote in 2001. So severe was the famine, wrote Hassan, that people "were forced to commit unheard of atrocities such as eating their own children and violating the sacred sanctity of the royal dead."

To Hassan, there is no doubt that abrupt climate change—sudden drought—led to devastatingly low flows of the Nile, that famine and poverty caused the collapse, and that "the Nile can be considered as the force which destroyed the civilization that it had nurtured." The Nile-fed Lake Faiyum, more than 200 feet deep, dried up entirely. As the Nile was to Egypt, climate was to the Nile. The sudden onset of low flows faithfully reflected suddenly dry conditions in the Nile headwater regions of Ethiopia and equatorial Africa. According to Hassan, a 1999 study from West Africa of dust in the Kajemarum Oasis of northern Nigeria recorded a "pronounced shift in atmospheric circulation" around 2150 BC that led to "less rainfall and a reduction of water flow in a vast area extending from Tibet to Italy." According to archaeologist Lauren Ristvet, the abrupt "aridification event" 4,200 years ago "has been recorded in 41 paleoclimate proxies in the Old World, from Kilimanjaro, Tanzania to Rajasthan, India, East Asia and the Pacific."

Beyond the Old World climate crashes, powerful and abrupt climate changes during the past 10,000 years have been detected in climate archives throughout the world, but most particularly in records from the energy-rich and water-rich lower latitudes that plot changes in the global distribution of rain and snow. As the French researcher Françoise Gasse of Université Aix-Marseille argues, the concept of a "fairly stable" Holocene may have more to do with the history of the science than with the history of Earth's climate. The idea came from polar ice and North Atlantic deep-sea core records, where fluctuations during the Holocene were much lower than the transition from the ice age, such as the Younger Dryas. "On the contrary," wrote Gasse in 2000, "dramatic hydrological changes have long been apparent in Africa, and appear as large in magnitude as their glacial counterparts. Dramatic changes in water resources have enormous consequences on human populations, generating famines, migrations, civilization foundations and collapses."

The last of the abrupt changes that paleoclimatologists say resembles the pattern of events during the ice age was a sudden but relatively brief episode of sharp cooling and drying that shows up in polar ice cores and North Atlantic sediments after the Younger Dryas, about 8,200 years ago. To paleoclimate researchers, this event has the earmarks of another ice sheet meltwater flood that changed the density balance of the thermohaline circulation and curbed the northerly flow of warm water into the North Atlantic. The 2002 report on abrupt change by the National Research Council noted that, coming more than 3,000 years after the Younger Dryas, this episode "punctuated a time when temperatures were similar or even slightly above more recent levels, demonstrating that warmth is no guarantee of climate stability." Measuring argon and nitrogen isotopes in air bubbles trapped in the Greenland ice cores as well as oxygen isotopes in the ice itself, Takuro Kobashi of Scripps Institution of Oceanography in San Diego and colleagues produced a striking profile of the event. In a period of only five years, temperatures plunged and the climate remained "locked" in a cold regime for about 60 years.

From Weiss's new archaeological perspective, this 6200 BC climate crash had a portentous impact on Neolithic farmers, who for 1,000 years had been engaged in rain-fed cereal agriculture in northern Mesopotamia, when drought suddenly struck. Some of them migrated downstream along the steep banks of the Tigris and Euphrates Rivers into the southern alluvial delta floodplain near the Persian Gulf, where streamflow could be managed and small levees could be breached to water their fields. Irrigation agriculture was twice as productive as dry farming in the region, Weiss noted in 2003, but the new, more labor-intensive methods also required new social organization. The climate crash 8,200 years ago, he said, "provided the natural force for Mesopotamian irrigation agriculture and surplus production that were essential for the earliest class-formation and urban life."

For the rest of the Holocene, that is, for the last 8,000 years, abrupt changes in precipitation patterns leave the trample marks of big boots through climate records close to where human societies have flourished and fallen. None was more dramatic than the changes 4,000 years ago that transformed the Sahara from a green, wet savannah of annual grasses and low shrubs that characterized the landscape since the beginning of the Holocene period into a vast desert, a dramatic change that has puzzled climate researchers for years. Subtle changes in the geometry of Earth's orbit are ultimately responsible—but why so sudden in the mid-Holocene, and why so severe?

As the great meteorologist Jule G. Charney had argued in 1975, the dynamic response of the atmosphere to vegetation loss begets the growth of a desert. In 1999, the German climate modeler Martin Claussen described how a powerful feedback mechanism provoked the most profound abrupt climate change of the last 6,000 years. Writing in *Geophysical Research Letters*, Claussen concluded that while the timing depends on global circumstances, "the abrupt desertification in North Africa during the mid-Holocene can be explained only in terms of internal, mainly regional, vegetation-atmosphere feedbacks in the climate system."

In 1991, a collaboration between anthropologist Izumi Shimada and tropical ice core specialist Lonnie G. Thompson secured one of the first solid connections between abrupt climate change and culture in the Western Hemisphere. Thompson's Quelccaya ice core from the high Andes of southern Peru allowed researchers to devise a detailed picture of the circumstances of abandonment and reformation of the Moche civilization in the sixth century. Centuries of imperial coastal Moche culture came to an end during a 30-year drought that was followed by severe El Niño flooding. The Moche capital was destroyed. Sand dunes filled the irrigation channels. There was widespread famine. Between AD 600 and AD 750, the succeeding Moche culture relocated its capital farther north and farther inland, near Andean rivers, where the supply of water was more dependable, and implemented new agricultural and architectural techniques.

In the Northern Hemisphere, nothing in history has equaled the disintegration of the great Maya civilization of the Yucatan Peninsula in what is now Mexico, Honduras, and Guatemala. For more than 3,000 years, from about 2000 BC, this society flourished through a "dark age" of European habitation and may have reached a peak population of 15 million. With their writing system and solar calendar, and their traditions of religion and nobility, the Maya engaged in astronomy and mathematics and practiced the architectural arts, building great temples and pyramids, palaces, and observatories.

Between the middle of the eighth century and the middle of the tenth century, the end came in several episodes of political consolidation, famine, disease, and disorder.

To explain the disaster, archaeologists rounded up their usual suspects, citing a weak economic base, a rigid political hierarchy, and palace intrigue by nobles vying for power. Some pinned the blame for the collapse on the effects of overpopulation and deforestation. Although problems such as these almost certainly affected the Maya at one time or another, 3,000 years of sustained existence would seem to be its own best argument against the notion that internal weaknesses caused the demise.

As archaeologists are inclined to emphasize, great episodes of history do not yield to simple explanations, to pat answers. In the case of the collapse of the Maya, in particular, arguments for the role of a changing climate have been hard-pressed to explain the complexity of this civilization's multistage, multifarious decline. The contraction began in the late eighth century with the abandonment of cities in the western rain forest; in the ninth century, in the central lowlands; and finally early in the tenth century, in the Maya heartland, where archaeologists estimate that by AD 930 95 percent of the population was gone.

Early pollen studies were inconclusive, mainly because researchers could not distinguish between the effects of human deforestation and those of a changing climate. In 1996, geologist David Hodell reported results of chemical analyses of a Yucatan saline lake's sediment record that pointed to "the first unambiguous evidence for climate drying between AD 800 and 1000." This period, which saw the Maya collapse, was the driest episode of the past 7,000 years in Hodell's record.

In 2003, the sediments of the Cariaco Basin, the best climate archive in the Tropics, yielded a more detailed picture of what happened during the Maya collapse. By a fortunate coincidence, the seasonal movement of the Intertropical Convergence Zone, the thunderstorm belt in which northern and southern trade winds come together, brings the same climate of dry winters and wet summers to both the Yucatan Peninsula and this ocean basin north of Venezuela. An international team led by German geologist Gerald H. Haug tracked rises and falls in the presence of titanium, which varies with patterns of rainfall and river runoff into the Cariaco Basin. Haug's results showed that within the general 150-year dry period identified by Hodell were three distinct spikes of extreme drought at AD 810, 860, and 910 that mirrored periods of Mayan social collapse.

"We suggest that the rapid expansion of Maya civilization from AD 550 to 750 during climatically favorably (relatively wet) times resulted in a population operating at the limits of the environment's carrying capacity, leaving Maya society especially vulnerable to multiyear droughts," the team reported in *Science.* Already in a dry period, already near the threshold, the Maya may have been pushed over the edge by distinct pulses of abrupt climate change that brought collapse first in one region and then another and, finally, early in the tenth century, in the heartland itself. Sadly, just two years later, according to the Cariaco Basin record, wet summers returned to Yucatan.

A similar pattern seems to have run its course south of the Equator with the disintegration of a southern Andean Peruvian culture that coincided with "abrupt, profound climate changes" detected in sediment cores taken from the bottom of Lake Titicaca, site of a large urban population center of the Tiwanaku civilization of a thousand years ago. The Tiwanaku engineered an elaborate cultivation system of raised fields that so improved agricultural production it stimulated a dense population growth that could not be sustained during dry periods. The story unfolds in research by geographer Michael W. Binford, anthropologist Alan L. Kolata, and others who constructed a timeline spanning 3,500 years in the southern Andes highlands. Dry-land agriculture came to the region of Lake Titicaca with the Chiripa culture about 1500 BC during a time of plentiful water, when the lake level rose 65 feet, and continued after the emergence of Tiwanaku about 400 BC. Raised-field cultivation was developed about AD 600; by AD 1000 it was the main source of local food production. The end came about AD 1150, a time when drought caused the lake level to fall as much as 55 feet. "The lack of water simply made the physical and biological functions of the fields impossible," the team wrote in 1997.

The period from the tenth century until the early fourteenth century was a relatively benign climate epoch of warm temperatures around the North Atlantic, at least, that saw a dramatic expansion of European civilization. Vineyards in sunny England began producing good wines. Fields of barley and wheat swayed in the mild breezes of Iceland, and Norse livestock farmers secured their settlements in Greenland. Whether the Medieval Warm Period was a global phenomenon is a question of continuing debate, although it certainly was not a uniformly benevolent time. Very different and less salubrious changes in the climate of North America were under way. Again, they were punctuated not by changes in temperature but by changes in the distribution of rain and snow. Tree ring studies and other innovative methods are continuing to bring the history of climate forward in time, although the science of the last thousand years is young and the picture is far from complete.

In a region where most native peoples did not erect enduring monuments or settle in large, permanent population centers, direct archaeological clues to the climate's role in human events are few. Whether the 26-year drought in the late thirteenth century in the American Southwest was the primary cause for the abandonment of many elaborate settlements by the Anasazi is a matter of continuing debate. Recent modeling studies suggest that a more complex set of circumstances led to abandonments as populations fell below a certain threshold, although there is no doubt about the impact of drought on the overall viability of desert agriculture. Nor is there any argument about the evidence for enormously long and intense droughts in the climate record of the last millennium in North America.

In 1994, geographer Scott Stine reported remarkably clear evidence for extremely long droughts in California's Sierra Nevada. Using radiocarbon dating, Stine ascertained the age of the wood in tree stumps he found rooted in the beds of modern mountain lakes, streams, and marshes. The mountain range had endured more than two centuries of drought from AD 892 to about AD 1112, Stine reported, and another 140 years of drought that began about AD 1209 and came to an end around AD 1350.

Across the plains and prairies of interior North America, vast sand dunes lie under a thin skin of vegetation that has been peeled back by climate more than once. The dunes have drifted in the winds of droughts that far exceeded the intensity and duration of anything in the modern record of the region, including the 1930s "Dust Bowl," the United States' iconic experience with climate change. In the last quarter of the thirteenth century the first of two "mega-droughts" of the past thousand years gave flight to the sands of the region during a time that coincides with Stine's Sierra Nevada drought and the abandonment of the Anasazi settlements of the Southwest.

In the 1540s, when Europe was experiencing what climate researchers call the Little Ice Age, a second mega-drought spread from northern Mexico. By the end of the century, it extended over much of the United States. This was the worst drought to hit North America in 500 years, and it was implicated in a recent study of a hemorrhagic fever epidemic that led to a colossal loss of native

Mexican lives in the second half of the sixteenth century. A 2002 study by Mexican epidemiologist Rodolfo Acuna-Soto and American tree ring researcher David W. Stahle pointed to climate-driven spikes in rodent populations as its probable cause. Epidemics in 1545 and 1576 killed an estimated 17 million people, more than twice the number of victims of the smallpox outbreak that the Spanish delivered to Mexico in 1520. "In absolute and relative terms the 1545 epidemic was one of the worst demographic catastrophes in human history," Acuna-Soto and Stahle wrote in the journal *Emerging Infectious Diseases*. This was the drought that greeted Sir Walter Raleigh's hapless colonists in 1587 as they stepped onto the beach at Roanoke Island.

New high-resolution climate profiles from tree rings, deep-sea and lake sediments, and ice cores are rewriting some interesting episodes of human history, although, as Gerald Haug and colleagues noted in their Maya study, technical and other problems remain to be solved. "Unfortunately," they wrote, "the limitations of temporal resolution and chronology in paleoclimatic records still present a major obstacle to the development of a globally meaningful view of Holocene climatic changes and their role in social change." Its own origins in geology and in polar ice have given paleoclimate research an odd, upside-down appearance. Ironically, because abrupt change was recognized first in ice age archives, researchers have more highly developed ideas about events of a distant past than they do about more recent human experience with climate.

Writing in *Science* in 1996, University of Arizona geoscientist Jonathan T. Overpeck, a leading Holocene climate researcher, observed that both the early Greenland ice core record and the modern instrumental climate record of the past 150 years have led to some seriously mistaken but "comforting conclusions" about the character of events of the past 10,000 years. In contrast to the ice age, Overpeck noted, "the current warm interglacial climate is often characterized as relatively stable, leaving the impression that climates of the future are likely to be more or less well behaved." And because it is the only period when changes were widely recorded by instruments, many people seem to believe that the record of the last century represents the full range of natural climate variations. With advances in research techniques has come a new picture.

"It is now clear that climate variability in many regions of the world, including Greenland, was significantly greater during the last 10,000 years than during the last 150 years," Overpeck wrote. Most of these warm-era events were smaller than ice age changes, but many were much bigger than any that instruments have recorded in the last 150 years. "More importantly," he wrote, "many of these past Holocene events appear to have been large enough that, if they were to recur in the future, they would have major impact on humans." David Stahle makes the point: "Year-in and year-out, over the long haul, drought extracts the most from humanity."

12

SURPRISES

Living in a climate of uncertainty

The Swiss geophysicist Hans Oeschger, one of the first researchers to lay eyes on the polar ice profile, recognized what the newly deduced pattern of natural climate behavior was likely to mean to the future. That disturbing vision inspired a sense of social responsibility in him that lasted the rest of his life. Oeschger realized that Earth's climate system does not always respond slowly to gradual "forcings" or modifying factors. Occasionally it crosses a threshold of some kind—and then it flips. The pace of change may have nothing to do with the pace of the forcing, Oeschger realized, and the climate's new equilibrium state, lasting a thousand years or more, is very different from the old one. At least two dozen of these gyrations—these Dansgaard-Oeschger events—punctuated the last 110,000 years of climate, and as he pondered their mysteries in the early 1980s, one conclusion he reached was that the nature of these changes is not friendly to humans. Like the gradual orbital influences that eventually pushed the climate system over a threshold into and out of ice ages, Oeschger warned, the gradual loading of greenhouse gases in the atmosphere could force the climate over another threshold into another state.

In his *Nature* obituary of Oeschger, Thomas F. Stocker, his successor at the University of Bern, wrote that as long ago as 1984, "Oeschger recognized the importance of ocean circulation in explaining abrupt climate change and used the physical analogy of a flip-flop system—triggered by small perturbations, the ocean circulation might switch from one circulation mode to another. Based on these insights, he was among the first to point out that the anthropogenic increase of CO_2 could represent such a perturbation. Although his early warnings were often greeted with disbelief. . . . "

Another key concept occurred to Oeschger as he pored over a job of editing some of his work with a colleague in the small office of an international climate research program in Bern. "He wrote with a Swiss accent, but his work also had a distinctive style and flare," Herman Zimmerman recalled of their work together in Bern in the 1990s. "We spent long hours editing out the accent, but keeping the flare. The term 'climate surprise' came from Hans' not being able to find quite the correct phrase in English, but we left it, because it actually described just the right thought—the abrupt, short-lived, climate changes of the glacial stages that had been unexpectedly found in the ice core records from Greenland."

To be sure, the confirmation of abrupt climate change in polar ice and other environmental archives was a big surprise to scientists in the 1990s. What really resonated with Oeschger and later researchers, however, was the sense that surprise inhabits the future in an unexpectedly important way—that the natural behavior of Earth's climate is both more dangerous in character and more uncertain than just about anyone had supposed. Most every theorist since Oeschger has employed variations on his "flip-flop" analogy and his concept of "climate surprise."

Depending on his audience, Richard B. Alley talks about a climate with dimmers and switches or compares abrupt change to a bungee jumper hanging over the side of a roller coaster car. *Boing! Boing!* About how "leaning slightly over the side of a canoe will cause only a small tilt, but leaning slightly more may roll you and the craft into the lake." About the drunkard who sleeps when left alone and staggers when awakened.

Wally Broecker, having conferred with Oeschger in Bern, was seeing the same writing on the same wall. In an influential 1987 commentary in *Nature*, Broecker wondered out loud if there would be "Unpleasant surprises in the greenhouse." With our "gigantic environmental experiment" of loading the atmosphere with carbon dioxide and other greenhouse gases, he wrote, "we play Russian roulette with climate, hoping that the future will hold no unpleasant surprises. No one knows what lies in the active chamber of the gun, but I am less optimistic about its contents than many."

Coming even before the so-called global warming debate, a dispiriting experience for many U.S. climate scientists in the 1990s, Broecker's commentary sounded a dark and pessimistic note. Records from polar ice and elsewhere indicate "that Earth's climate does not respond to forcing in a smooth and gradual way," he wrote, and the consequences of forcing from the buildup of greenhouse gases "are potentially quite serious." Our food supply and many species of wildlife could be at risk. "To date, we have dealt with this problem as if its effects would come in the distant future and so gradually that we could easily cope with them," he wrote. "This is certainly a possibility, but I believe that there is an equal possibility that they will arrive suddenly and dramatically." Researchers didn't know enough to make reliable predictions, and indeed, "reliable modeling may never be possible." Even with a great intensification of effort, he wrote, "I fear that the effects of the rise in concentration of the greenhouse gases will come largely as surprises."

In a variety of forums and formats, Broecker tried to jolt scientists and citizens out of what he saw as a dangerous complacency on the subject, warning in *Natural History* magazine in 1987, for example, that rather than "treating it as a cocktail hour curiosity, we must view it as a threat to human beings and wildlife that can be resolved only by serious study over many decades." Ten years later, in *GSA Today*, a journal of the Geological Society of America, Broecker asked: "Will Our Ride into the Greenhouse Future Be a Smooth One?" He warned that a doubling of the carbon dioxide concentration of the atmosphere would come as the planet is asked to feed double the human population during the next century. He reviewed the evidence from the Greenland ice and elsewhere, placing new emphasis on the possibility of water vapor and conditions in the Tropics triggering abrupt change.

"Whatever pushed Earth's climate didn't lead to smooth changes, but rather to jumps from one state of operation to another," he wrote. "So the question naturally arises, What is the probability that through adding CO_2 we will cause the climate system to jump to one of its alternate modes of operation?" We don't have any idea, he declared, although we do know that warming temperatures can reduce the density of polar surface water and cut off the circulation that carries tropical warmth through the North Atlantic. "So we're entering dangerous territory and provoking an ornery beast," he wrote. "Our climate system has proven that it can do very strange things.

"Since we've only recently become aware of this capability, there's nothing concrete that we can say about the implications," he conceded, although he saw in this uncertainty a strength to the argument. "This discovery certainly gives us even more reason to be prudent about what we do, though. We must prepare for the future by learning more about our changeable climate system, and we must create the wherewithal to respond if the CO_2-induced climate changes are large, or, worse yet, if they come abruptly, changing agricultural conditions across the entire planet. We must think all this through. Even if there is only a 1 percent probability that such a change might occur during the next 100 years, its impact would be sufficiently catastrophic that the mere possibility warrants a lot of preparation."

In the 1990s, the danger of greater risk from essentially unpredictable natural forces was about the last piece of information that policy makers wanted to know about this troublesome subject. And so it seems that most of them did not hear it, or at least did not hear it in a way that could become part of the cantankerous political dialogue of the times. In any case, there were caveats by the carload. No one could be certain, of course, but the abrupt changes detected in Greenland ice cores looked to most researchers like ice age events, whereas the Holocene (last 10,000 years) climate seemed to have been blessedly stable. Whatever the case, uncertainty is inherent in surprise, and uncertainty simply is not welcome in the political arena.

The subject of natural variation and the uncertain potential for abrupt change seldom came up in the hotly partisan 1990s U.S. political debate about "global warming" and the Kyoto Protocol. The closer Hans Oeschger had looked at the problem, the more seriously he had taken science's social responsibility. Colleagues remember him saying, "The worst for me, would be if there were serious changes in the next five to ten years and we scientists . . . did not have the courage to point out these dangerous developments early on." But climate scientists who took up Oeschger's challenge found themselves in an arena that was not congenial to science or receptive to their qualified trains of thought.

In the greater marketplace of ideas, transactions are brokered by people who often are immaculate of science and who have their own, more certain ideas about the future. Like other marketplaces, it is a realm in which the point of view's viability can be influenced by the amount of money its adherents have to spend. It is nothing if not democratic, such a marketplace; one idea is just as good as another and every story has two sides.

Climate scientists were treated in the mass media not as courageous researchers performing acts of social responsibility, but as just another interest group—and a poorly financed one at that. More often than not, they found themselves arguing with industry lobbyists, who were given equal billing, in circumstances designed by television producers and editors to elicit the greatest possible conflict. So disagreements were frequently passionate and often devoid of science. Before long, the subject was just another "hot button" topic, an article of political faith that had nothing to do with climate. Global warming was something one either "believed in" or most definitely did not. Even among believers, most often it was seen as a gradually worsening problem of uncertain consequences that would manifest themselves in a manageable way over time.

The United Nations-sponsored Intergovernmental Panel on Climate Change, which began raising the possibility of "climate surprises" from feedbacks in the system in 1991, was still, more than 10 years later, composing graphical representations of the future that depicted congenially smooth curves of gradual change as far out as computer models could see. And more than a decade after the Greenland ice cores confirmed abrupt change, virtually all political discourse in the United States remained devoted to the idea that climate conforms to the rules of a well-behaved, linear system.

Climate crashes such as those that punctuate the paleoclimate archives are said to be high-impact, low-probability scenarios, like asteroid impacts or other catastrophes, although even this assertion raises questions that climate scientists cannot easily manage. Who really knows? Now they can say with certainty that thresholds exist in the system and that crossing a threshold could suddenly pitch the climate into a new mode of operation that societies could find difficult to adapt to—or worse. But who can say how close or how distant the next abrupt change is?

By the turn of the new millennium, finally the idea of abrupt change had reached some kind of critical mass among climate scientists. In a coming of age for the science, the National Research Council appointed a special committee to conduct a comprehensive review of the subject. Some of the brightest climate scientists in the business were asked to identify "critical knowledge gaps" and to recommend a research strategy. In 2002, the NRC committee issued its report *Abrupt Climate Change: Inevitable Surprises.*

"We do not yet understand abrupt climate changes well enough to predict them," Richard Alley, who chaired the NRC committee, wrote in a preface to the report. "The models used to project future climate changes and their impacts are not especially good at simulating the size, speed, and extent of the past changes, casting uncertainties on assessments of potential future changes. Thus, it is likely that climate surprises await us."

A problem that Alley and other paleoclimate scientists refer to as the "insensitivity of models" or the "model-data gap" sounds like a technical issue but really is more fundamental. It means that the models are unable to reproduce accurately the numerous episodes of abrupt change that show up clearly in many environmental archives around the world. The reasons for this failure are not yet known, but the implications are plain enough. Until these highly sophisticated numerical representations of Earth's climate system—running on the world's most powerful computers—are able to get the past right, what reason is there to believe they can get the future right?

This is a divide that separates two general trains of thought about climate's behavior in the future. How good are the models? How complete is the theory of climate? It is such questions that most divide climatologists' opinions—not the fact that changes are coming or that greenhouse gases will cause them, or the basic chemistry, so much as the underlying physics: the steepness and slipperiness of the slope.

On one side is the well-established idea that the models, although imperfect, still best represent the behavior of warm-era climate—they get the *recent* past right—so they are the best guides to the future. This line of thought assumes that climate will continue to behave as it has much of the time—a small push usually leads to a small response—and the record of the recent past proves the point. On the other side is the upstart idea, born out of the Greenland ice cores, that climate is a precariously balanced nonlinear system that lurches between very different states of coldness, dryness, wetness, and warmth. This is the meaning of the "climate surprise" warning flag that abrupt-change researchers are waving—small changes can provoke big responses, and the paleoclimate record proves the point.

"Although abrupt climate changes have shocked ecosystems and societies over the last few millennia," the NRC committee observed, "these changes have not been as dramatic as those that occurred during the last ice age. It is probably no coincidence that stability of the climate increased when ice-sheet size and atmospheric carbon dioxide concentration largely leveled off at the end of the ice age."

Could the continuing buildup of greenhouse gases in the atmosphere during the modern industrial era provoke climate to change abruptly? Yes, the NRC committee concluded, it could. Abrupt change could happen any time in a chaotic system such as Earth's climate, said the report, but the "existence of a forcing greatly increases the number of possible mechanisms. Furthermore, the more rapid the forcing, the more likely it is that the resulting change will be abrupt on the timescale of human economies or global ecosystems."

The paleoclimate record shows that "the abrupt climate changes of the past were especially prominent when orbital processes were forcing the climate to change most rapidly during the cooling into and warming out of the ice age, consistent with the results from modeling that forcing of climate increases the possibility of crossing thresholds that trigger abrupt change," the committee wrote. "Given our understanding of the climate system and the mechanisms involved in abrupt climate change, this committee concludes that human activities could trigger abrupt climate change. Impacts cannot be predicted because current knowledge is limited," but there are many possibilities. Such well-known modes of annual variation such as El Niño or the North Atlantic Oscillation could change their behavior. Droughts could become more frequent or widespread or show up in unexpected places. And the circulation of the North Atlantic could slow down or speed up.

Across different timescales, almost every important climate change, from the advance and retreat of glaciers to the comings and goings of El Niño, is driven by closely coupled feedback loops between conditions in the oceans and the atmosphere. Winds control ocean currents that control sea surface temperatures that control atmospheric pressure differences that control winds—and so on. The North Atlantic Oscillation is another ocean-atmosphere feedback loop that steers the tracks of storms over much of the Northern Hemisphere as it waxes and wanes. In the new context of abrupt change, of a chaotic system, is it really possible for climate science to untangle all of the linkages, to categorically ascribe some changes to human-induced greenhouse gas loading and others to natural variation and to predict the future?

The modern climate system slips in and out of preferred modes of behavior such as a stronger or weaker North Atlantic Oscillation or Pacific Ocean patterns that seem to alter the intensity and frequency of the El Niño-Southern Oscillation. These various changes in "decadal variability" not only have powerful and far-ranging impacts on global precipitation, drought, and critical monsoon patterns, but also may affect one another. If the system is nearing a threshold of abrupt change, could one of these natural shifts push it over the edge? It is a measure of the nonlinear nature of the system, this Gordian knot of climate, that potential upstream causes of change and potential downstream effects cannot readily be distinguished from one another.

Perhaps it is no wonder that economists and social scientists have only begun to take up the idea of abrupt change and assess its potential to alter the economic well-being and stability of modern societies. In 2003, Alley and other members of the NRC committee noted in *Science*: "Although there is a substantial body of research on the ecological and societal impacts of climate change, virtually all research has relied on scenarios with slow and gradual changes. In part, this focus reflects how recently the existence of abrupt climate changes gained widespread recognition, and how difficult it has been to generate appropriate scenarios of abrupt climate change for impacts assessments." In addition, the scientists noted, the United Nations Framework Convention on Climate Change "has focused attention on anthropogenic forcing, whereas abrupt climate change is a broader subject covering natural as well as human causes."

An increasing number of studies that link economic and climate models "indicate that many sectors of the economy can adapt to gradual climate changes over the coming decades. But this research sheds little light on the impacts of abrupt climate changes, particularly where these involve major changes in precipitation and water availability over periods as short as a decade," and there is "virtually no linked research on abrupt climate change." While economic estimates based on the gradual warming scenario indicate that climate change could be slowed by "modest but increasing emissions reductions and carbon taxes . . . efficiently avoiding abrupt change may involve much larger abatement costs."

Even in the United Kingdom and northwestern Europe, a region that is in the cross-hairs of the dominant scenario for abrupt change, economists and social scientists have only begun to study the potential impact of a slowdown or collapse of thermohaline circulation in the North Atlantic.

In 2002, a preliminary report by the UK's Natural Environmental Research Council noted that a "sudden strong cooling could be catastrophic for agriculture, fisheries, industry and housing," although another UK study noted that "neither the probability and timing of a major ocean circulation change nor its impacts can be predicted with confidence yet."

Ominous signs of change are being detected in the North Atlantic. In 2002, in *Nature*, Robert Dickson of the UK's Centre for Environment, Fisheries, and Aquaculture Science described "a widespread, sustained, rapid and surprisingly uniform freshening of the deep and abyssal North Atlantic, south of the Greenland-Scotland Ridge, over the past four decades." It was the greatest change in oceanography ever recorded in the modern era, although still smaller than ice age changes recorded in seafloor sediment cores. Alley and the NRC committee noted in *Science* this "invasion of low-salinity deep waters that spread over the entire subpolar North Atlantic Ocean and the seas between Greenland and Europe in just the regions critical for abrupt shifts in thermohaline circulation, which has been implicated in many abrupt climate-change events in the past." In 2003, also in *Nature*, physical oceanographer Ruth Curry reported marked salinity changes in surface waters through the western basins of the North Atlantic between the 1950s and 1990s. A "systematic freshening" is occurring near both poles,

Curry wrote, and a "large increase of salinity" is being detected at low latitudes. These shifts are worldwide, she wrote, and "suggest links to global warming and possible changes in the hydrologic cycle of the Earth." In spring 2004, in *Science*, oceanographers Sirpa Häkkinen and Peter Rhines reported satellite measurements detecting a slowing of the North Atlantic current known as the subpolar gyre.

Other researchers are monitoring critical conditions in the polar ice sheets, where abrupt changes could lead to global sea level rise. In 2004, two reports in the journal *Geophysical Research Letters* reported rapid changes in West Antarctica following the 2002 collapse of the Larsen B ice shelf. Using satellite images, the two teams reported glacier acceleration and thinning and accelerated ice discharge from the Antarctic Peninsula. Other satellite reports cited accelerated melting of the Greenland ice sheet and a rapid decline in Arctic sea ice. In the fall of 2004, a multinational comprehensive study of the Arctic reported that the region was experiencing "some of the most rapid and severe climate change on Earth."

In "The Perfect Ocean for Drought," as researchers Martin Hoerling and Arun Kumar reported in *Science* in 2003, cold sea surface temperatures cover much of the eastern Pacific and warm surface temperatures prevail in the western tropical Pacific and Indian Oceans. Simulations by these climate modelers and other researchers have linked these ocean temperature patterns to drought conditions across the globe. The persistence of this pattern from 1998 to 2002 accounts for four years of drought across North America, southern Europe, and southwestern Asia. And still, through 2004, unusually warm temperatures prevailed in the Indo-western Pacific Tropics, and the U.S. Southwest and central-southwest Asia continued to endure drought. Some researchers saw the signature of global warming in the pattern. Paleoclimatologist Jonathan Overpeck called it the biggest drought in the United States since records have been kept and wondered if "we've pushed the climate system over a threshold."

Overpeck's and other researchers' focus on the Pacific Ocean is part of a noticeable shift in the attention of climate scientists toward events in the Tropics as a region from which abrupt changes are driven. While the North Atlantic may have been the critical center of action in a world dominated by ice sheets (the argument is not settled), events in the Tropics seem especially potent in a world dominated by warming. The tropical specialist Raymond T. Pierrehumbert heralded the new perspective in the *Proceedings of the National Academy of Sciences* in 2000: "Climate Change and the Tropical Pacific: The Sleeping Dragon Wakes."

The failure of computer models that simulate shutdowns in Atlantic Ocean circulation to accurately reproduce the effects that these abrupt changes left behind in the climate record may mean that the trigger for such events is to be found elsewhere in the system. Moreover, "given that we don't know what accounts for the unusual stability of the recent climate, we don't know what it would take to break it," Pierrehumbert wrote. "This thought is unsettling in a world seemingly committed to substantial warming from anthropogenic CO_2 increases in the next 2 centuries."

Pierrehumbert identified "an intimate link between events on the margin of Antarctica, and the tropical Pacific processes that govern both El Niño and Pacific oceanic heat transport." Thus, events in and around Antarctica, along with changes in the Tropics, bear watching.

"Whatever the world ocean does, the tropics may respond with some surprising reorganizations of convection," he wrote. There is "nothing inevitable about the present configuration" of the tropical Pacific—of a "warm pool" of water in the far western Pacific, the "cold tongue" of sea surface temperatures extending out toward the central ocean from the equatorial shore of South America, and a band of towering convective storms known as the Intertropical Convergence Zone that snakes north and south of the Equator with the seasons. It's just possible that the whole thing could break down—that the easterly trade winds could be replaced by a "westerly superrotation" leading to far-reaching changes in climate. It's an "exotic possibility," to be sure, and there is no evidence that such a state ever existed in Earth's climate. But the same might be said of the near future. A state that includes especially high concentrations of carbon dioxide in the atmosphere at the same time that ice covers the poles is "different from any that has ever before held sway on the planet," he wrote. "If one is tugging on the dragon's tail with little notion of how much agitation is required to wake him, one must be prepared for the unexpected."

In 1997, Wally Broecker wrote in *GSA Today* that he had been "humbled" by his lifetime study of Earth's climate, a circumstance that may have surprised a few graduate students at Columbia University. "I'm convinced that we have greatly underestimated the complexity of this system," he wrote. "The importance of obscure phenomena, ranging from those that control the size of raindrops to those that control the amount of water pouring into the deep sea from the shelves of the Antarctic continent, makes reliable modeling very difficult, if not impossible. If we're going to predict the future, we have to achieve a much greater understanding of these small-scale processes that together generate large-scale effects."

The more likely circumstance, as Broecker was among the first to suggest, is that we are not going to predict the future with any degree of confidence—that climate surprises are inevitable. As Alley and the NRC Committee on Abrupt Climate Change, writing in *Science* in 2003, put the case: "The difficulty of identifying and quantifying all possible causes of abrupt climate change, and the lack of predictability near thresholds, imply that abrupt climate change will always be accompanied by more uncertainty than will gradual climate change."

In 2002, in a lecture to the American Geophysical Union meeting in San Francisco, Alley reviewed the performance of the major climate models on which the Intergovernmental Panel on Climate Change has based its forecasts for a globally warming Earth. When it comes to simulating the past, across the board the models underestimate the changes that are known to have taken place. "On average, the models got two-thirds of what happened," he said. "It's not a cold bias, it's not a warm bias. It's an insensitivity to changed boundary conditions . . . how sensitive the model is when you change things."

"The least astonishing hypothesis that I get from this is that either the future warming projections are accurate or they have underestimated what we face in the future," he said. "There's a lot of paleoclimate work to be done here to test this hypothesis," but if climate models are systematically underestimating reality, then the more radical-looking "high side" of temperature projections may be more accurate than the conservative-looking "low side."

Not only are policy makers presented with what are likely to be overly optimistic expectations of the future, they are presented a profile with contours that look nothing like the record of climates past. The changes of the past are single lines or narrow bands that represent real data, whereas the projections of the future prepared for policy makers are large smoothed curves that represent the average results of many computer model simulations and a wide range of possibilities. "This tendency of the policy maker to see a smooth curve has to be really disturbing," said Alley. "Because whatever it's going to do, if it's smooth, we're going to be really surprised. It's going to stagger, it's going to jump. What happens regionally is not going to happen globally. And we really, I think, need to look at how variable it will be. What's possible in the system, and where does it go?"

The news from Greenland, unfortunately, is that much more is possible in the climate system than anyone would have supposed. Changes can be big and fast and potentially dangerous to societies that are heavily invested in stability and resistant to adaptation. In the event of an abrupt change—a climate surprise—political arguments probably will no longer be about industrial emission controls, their fairness or economic viability. More urgent political and economic problems will command the attention of nations around the globe. In the event, it is in nature's power to so change our world that even such basic questions as cause—whether the crash came naturally or not—could seem sadly beside the point.

FURTHER READING

Books

Abrupt Climate Change: Inevitable Surprises, National Research Council, National Academy Press (2002).

The Discovery of Global Warming, by Spencer R. Weart, Harvard University Press (2003).

Ice Ages: Solving the Mystery, by John Imbrie and Katherine Palmer Imbrie, Harvard University Press (1986).

The Ice Chronicles, by Paul A. Mayewski and Frank White, University Press of New England (2002).

The Little Ice Age: How Climate Made History, 1300-1850, by Brian M. Fagan, Basic Books (2001).

The Long Summer: How Climate Changed Civilization, by Brian M. Fagan, Basic Books (2004).

The Two-Mile Time Machine, by Richard B. Alley, Princeton University Press (2000).

Articles

"Abrupt Climate Change," by Richard B. Alley, *Scientific American,* November 2004.

"Annals of Science: Ice Memory," by Elizabeth Kolbert, *New Yorker,* January 7, 2002.

"Chaotic Climate," by Wallace S. Broecker, *Scientific American,* November 1995.

"The Discovery of Rapid Climate Change," by Spencer Weart, *Physics Today,* August 2003.

"Feel the Pulse," by Fred Pearce, *New Scientist,* September 2, 2000.

"The Great Climate Flip-Flop," by William H. Calvin, *Atlantic Monthly,* January 1998.

"Greenland Ice Cores—Frozen in Time," by Richard B. Alley and Michael L. Bender, *Scientific American,* February 1998.

"Ice-Cold in Paris," by Stefan Rahmstorf, *New Scientist,* February 8, 1997.

"In Deep Water," by Robert Kunzig, *Discover,* December 1996.

"Learning from Polar Ice Core Research," by Deborah Schoen, *Environmental Science & Technology,* April 1, 1999.

"The New Ice Age," by Brad Lemley, *Discover,* September 2002.

"The Pentagon's Weather Nightmare," by David Stipp, *Fortune,* January 26, 2004.

"Perilous Waters," by Charles W. Petit, *U.S. News & World Report*, April 1, 2002.

"Rapid Climate Change," by Kendrick Taylor, *American Scientist*, July-August 1999.

ACKNOWLEDGMENTS

For their guidance and patient assistance at various stages of this project, I am indebted to many earth scientists, most particularly to Chester C. Langway Jr., Richard B. Alley, Kendrick Taylor, Lisa K. Barlow, Jonathan T. Overpeck, James W. C. White, James P. Kennett, John M. Wallace, Wallace S. Broecker, Gerard C. Bond, Gavin A. Schmidt, Spencer R. Weart, and Ronald A. Liston.

Thanks also to Jeffrey Robbins, senior editor at Joseph Henry Press, for his valuable help.

INDEX

A

F

G

H

I

J

K

L

O

P

T

U

V

W

Y

Z